November 2005

To Mike

Good luck with

Andrea & career -

Ross Wynn

Talking to the Sick

A Clinician's Guide to Effective Communication

FIRST EDITION

Richard J. Ugoretz, M.D.
Clinical Professor of Medicine
University of California, San Diego
School of Medicine

Talking To The Sick
ISBN: 978-0-557-09865-1

CONTENTS

Acknowledgments

I wish to acknowledge the distinguished mentors who modeled the clinical skills that guided all stages of my career in medicine. Remarkable good fortune allowed me to benefit from the influence of Dr. Samuel Asper, Dr. Phillip Tumulty, Dr. C. Lockard Conley, Dr. William Morgan and Dr. Eugene Braunwald. Innumerable patients have also inspired me with their insights, generosity and fortitude, even while facing adversity.

Special thanks are due to my editor, Sheila Berg, without whose patience and invaluable assistance it would not have been possible to complete this project.

Introduction

Consciously or unconsciously we adopt different patterns of speech depending on the identity of the listener and the setting in which we are speaking. This book concerns two specific groups of speakers and listeners - groups with much to gain when communication is successful. The speakers are doctors, who are members of their group by choice. The listeners are the sick, permanent or temporary, but always involuntary members of their group. At one time or another, life will grant nearly all of us membership in this group.

My career as an oncologist spanned an era of unprecedented growth in medical science - technical advances in diagnosis, treatment and understanding of human biology that were unimaginable when I entered medical school. Looking back on my own experience as a doctor, I realize that many of my successful encounters with patients and some of my failures were based, not on the influence of science, but on whether I was able to talk effectively to sick people.

Physicians in developed countries now practice in an environment where biologic science and medical technology occupy a central, at times dominant, role in the profession. In this environment reliance on technology can easily overshadow the human qualities that shape relationships between doctors and patients. Paradoxically, in spite of the impressive array of technologic assets available to them, modern physicians may be at a serious disadvantage in respect to communication with patients compared to their forebears. The crucial asset lacking in modern medicine is time.

Time that was once spent listening to and talking to the sick is now consumed by myriad competing activities; ordering, scheduling, and reviewing the results of diagnostic tests, communicating with consultants, specialists and paramedical personnel and dealing with the economics of medical practice. Innumerable third parties - insurance companies, health maintenance organizations, professional societies and government agencies - also require the doctor's attention. In addition, modern transportation, fax machines, cell phones, digital communication and the Internet have widened the scope and vastly accelerated the pace of medical practice.

The physical environment in which medicine is practiced has also changed. The severely ill now spend far more time in the hospital than the home, especially at the end of life. The nineteenth-century medical office often looked more like a study than a clinic. Today much office practice takes place in less reassuring surroundings, spartan at best, cramped and sterile at worst. A nineteenth-century visit with a doctor might well have occurred in a patient's home. If in the physician's office, doctor and patient might have been seated on upholstered furniture surrounded by bookshelves.

Following some conversation the patient might step behind a screen, disrobe and put on a gown. There might be a simple examining table in the same room, and after the examination the patient would dress, return to his or her chair and have further conversation with the doctor. There would be few, if any, laboratory tests or x-rays to explain, let alone computer tomography, MRI scans or radionuclide scans to schedule or discuss.

Contrast this scene with the standard twenty-first-century examining room where the décor is often limited to a few drug company-sponsored exhibits and a computer. The patient may have been led in by an assistant, handed a paper gown and asked to wait. Severely pressed for time, the doctor enters, sits at a computer or remains standing, asks a number of questions and proceeds quickly to an examination. What time remains must address test results, questions, advice and instructions for prescriptions. Constrained by closely scheduled visits and frequent interruptions, the doctor may be out the door in less than fifteen minutes. Even the most highly motivated contemporary doctor will be challenged to communicate well in this setting.

Effective communication has not been made less critical to medical practice by the advance of technology. Sick people still have the need to be listened to and to have their doctors speak to them in a way that is at once responsive, lucid, informative, and reassuring. The sick are still likely to be frightened, angry, depressed or misinformed. Although they might share personal or cultural backgrounds with their doctors, more often than in the past, doctor and patient come from vastly different backgrounds. These factors, taken together with the increased complexity of diagnosis and treatment, require doctors to have at least as much skill and use at least as much care in talking to the sick as they did in the past.

During my years in practice sick people often told me about their experiences with other doctors. Too many of them described important things that doctors failed to say or should have left unsaid or expressed more effectively. I always tried to view these descriptions with open-minded neutrality. After all, they were always one-sided and sometimes given in a background of anger or disappointment. However, I have made enough mistakes of my own in talking to patients and witnessed enough similar mistakes made by other doctors to convince me that many patients describe their experiences accurately.

What I have to say is neither novel nor even original. The guiding principles for talking to the sick were elaborated by ancient physicians beginning with the Hippocratic writings. In the more recent past gifted clinicians and teachers of great skill, including Dr. Philip Tumulty and Drs. William Morgan and George Engel, have addressed this subject in texts that, although out of print, are still relevant (Tumulty 1973; Morgan and Engel 1969).

Medical school curricula have been revised to include instruction in communication skills. Continuing medical education programs have aimed to improve the way practicing physicians talk to patients. However, published studies that have attempted to evaluate the success of these efforts report, at best, mixed findings as to their success (Brown et al. 1999; Yedidia et al. 2003; Merckaert et al. 2005; Hobma et al. 2006). Most troubling, reports of patient dissatisfaction with doctor-patient communication seem to be increasing rather than decreasing (Braddock and Snyder 2005).

My goal in writing this book is to describe strategies that lead to more effective communication between doctor and patient, especially in contemporary American medical practice and the setting of serious illness. Doctors have, not only the highest responsibility for effective communication with the sick, but also the most to lose, after patients themselves, when communication is unsuccessful. As Tumulty wrote in 1973, "What the scalpel is to the surgeon, words are to the clinician. When he uses them effectively, his patients do well. If not, the results may be disastrous." (Tumulty 1973, p. 4).

CHAPTER ONE GENERAL PRINCIPLES

D octors make unconscious choices when they approach a sick person that have a profound effect on how that person perceives the doctor's attitudes and intentions. Some of these choices involve unspoken communication. The principles discussed here, though self-evident, deserve emphasis because they are so often ignored in the conduct of modern medicine.

The New Patient Visit

Introduce yourself when you meet a new patient. Taking the initiative in saying, "Hello, my name is Dr. _____," not only announces your identity but also conveys the fact that you hold a measure of respect for your patient. In the same vein, give some thought to the way in which you address your patient. Your decision whether to use Mr./Ms./Miss, a professional title, a surname or only a first name influences the impression you make.

Several published studies have indicated that many patients are comfortable being addressed by their first names

(Gillette et al. 1992; Makoul et al. 2007; Amer and Fischer 2009). My advice is not to assume this is the case when you meet a patient for the first time. After all, you have probably just used the title Doctor when you introduced yourself. It is safer to assume that your patient would appreciate the use of a formal mode of address at your first meeting. It is easy to change to a first name when invited to do so. To assume that is what the patient wishes without being asked misses an easy opportunity to demonstrate respect. One of the important goals in a first encounter is to establish a foundation of respect that will, it is hoped, become mutual.

Occasionally you will meet a patient who takes the initiative and asks if he or she may address you by your first name or, more significantly, does so without asking. This may simply be an expression of friendliness. But it has been my experience that, when it occurs early in the course of an initial medical visit, it is a clue that suggests your patient has some insecurity about the power balance in your relationship. When a new patient seizes the initiative in asking for or using your first name, especially when you have been addressing the patient as, for example, "Mr. Jones," decide whether you will be entirely comfortable being addressed in that way. If not, my advice is to smile and say, "Let's leave it at Dr._____ for now." Though it may seem rather stuffy to take this position, it is preferable to setting the precedent for a level of familiarity that you later wish you had not.

Nonverbal physician behavior can influence communication in medical practice. One simple behavior requires no extra time, special skill, training or experience. It does require one vital piece of equipment that, surprisingly, is often missing in hospital rooms and some office examining rooms: a chair. It has been shown experimentally that a patient perceives that he or she has spent more time with a physician when the physician is seated than when he or she is standing (Johnson et al. 2008). Beyond the benefit of this positive perception, being seated sends an unspoken message that the patient is the focus of the physician's attention and his or her first priority. Instead of looking down at a seated or recumbent patient, a seated physician can look at a patient eye to eye.

Here too, there is at least the potential for sending the unspoken message, "I am a partner in this effort, I am willing to listen, and I am not asserting superiority." How frequently have you heard the complaint that none of these conditions applies to a physician? Studies have also shown that seated physicians are perceived as more compassionate by patients (Strasser et al. 2005; Bruera et al. 2007). Simply finding a chair or stool or, if there is no other choice, using the edge of a bed will give you a head start on avoiding the negative impact of the unintended messages implied by standing. It is always wise to ask permission if you must sit on a patient's bed. The hospital bed is the patient's territory in an environment where he or she has already sacrificed a good deal of control. This thoughtful gesture makes it clear that you care about the patient's autonomy, takes no more than a few words and will pay dividends in your relationship.

In a comprehensive review of the literature on doctor-patient communication (Beck et al. 2001), the following behaviors were among those found to have significant associations with favorable outcomes: leaning toward the patient, facing the patient and uncrossed arms. Excessive social touching, and the opposites of the previous behaviors, were associated with unfavorable outcomes. It should be noted that only eight published studies met the reviewer's criteria for reliability, and the sample sizes for all the studies were small.

In the interest of saving time it is common to have patients undress and don a gown, usually paper, before the doctor arrives. This might be justified in the setting of a follow-up office visit, the emergency room or a scheduled procedure, but it is a bad idea in the case of an initial visit. Although a controlled study concluded that a majority of patients did not object to the practice (Meit et al. 1997), there is much to be gained by having your first meeting occur with a fully clothed new patient.

People come to see a doctor in a frame of mind that can range from perfectly calm through mild anticipation to outright terror. You have no way of knowing in advance where your patient is on this spectrum. It makes sense to create a setting that is as relaxed and comfortable as possible to allow for those who are at the anxious end of the spectrum. Waiting to see a doctor while dressed in a paper gown in an examination room fosters one of several unhelpful states of mind for the anxious patient, including vulnerability, increased anxiety and irritation. These states of mind increase the probability that your patient will forget what he or she wanted to tell you or, conversely, set the stage for the patient's intensely pressured effort to tell you everything all at once.

Slow Down

You have so much to say - questions, answers, test results, explanations, advice and so on - and so little time to say it. There is a great temptation to talk too fast. Nevertheless make a conscious effort to slow down. A highly valued colleague and I never tired of remembering the hilarious comedy routine by Bob Elliot and Ray Goulding in which they interview a maddeningly slow-talking member of the fictional Slow Talkers Association. Although this is a caricature, it is not a bad model for doctors to use to resist the impulse to talk too fast.

Think of the last time you had a conversation with someone who habitually talks too fast. It was probably not a pleasant experience. Furthermore, if your own speech is so rapid that your words are misunderstood or missed completely by your patient, the content matters little. The message did not get through. Speaking slowly contributes to the impression that you care about your listener and that you have given careful thought to what you say. All these observations are doubly true with the elderly and those whose first language is not English.

Racing through your message leaves a patient at least confused and, occasionally, irritated. If the patient has to ask you to repeat what you said you have lost whatever time you saved by talking fast. Worse, if for some reason the patient is unable to understand you and does not ask you to repeat what you said, the stage is set for disappointment. Pauses are very helpful, especially when the subject is complex. If these reasons for talking slowly are not sufficient, remember that your chances of uttering an awkward phrase or making a harmful slip of the tongue increase as the speed of your speech increases. Yes, you are needed elsewhere. Yes, you are accustomed to rapid, abbreviated speech with your colleagues. Yes, the HMO allows only ten minutes per visit. Talk slower anyway. It will pay off.

In a Manner of Speaking

The prevailing atmosphere in medical school and house staff training, characterized by competition, heavy workloads and the need to assimilate huge amounts of information, fosters speech that is not only rapid but also authoritarian, coldly precise and detached. This is not ideal preparation for talking to sick people who need and appreciate warmth, empathy and personal connection. The importance of several communication techniques that respond to these needs can hardly be overstated.

These techniques include orienting statements - telling patients what to expect in a visit or procedure, use of humor, solicitation of opinions, invitations to ask questions and efforts to confirm understanding. Primary care physicians who used such techniques frequently were found to have fewer malpractice claims than those who used them less often (Levinson et al. 1997).

The subtler influence of tone of voice is at least equally important. In tense situations it is all too easy for a doctor's feelings to color the tone of his or her speech. In a study focusing on malpractice claims, the voice tones of surgeons speaking at the beginning and end of patient visits were analyzed for the variables of warmth, hostility, dominance and anxiety without reference to the content of their speech (Ambady et al. 2002). Remarkably, the authors found that they could predict the surgeon's malpractice claims status based on the levels of these vocal qualities. If an extreme benchmark such as malpractice claims status is favorably influenced by better ways of speaking, surely overall patient satisfaction must be as well.

How can a physician ensure favorable qualities of speech and guard against unfavorable ones? Two practices should be considered: one involves listening, the other speaking. First, listen critically to your own language and tone of voice and consider whether you are talking to your patient the way you would hope to be spoken to in a similar encounter.

Second, use language in a way that indicates that you think of your patient as a person rather than a case of illness. An effective way to do this is to take advantage of any chance to inquire about a patient's background and personal life before tense situations arise. A study of physician communication found that surgeons' failure to show interest in their patients equaled failure to explain the medical condition as the most common reason for patients to choose not to recommend their surgeon to friends or family (McLafferty et al. 2006).

Most people appreciate an expression of interest in the qualities that make them unique. If done with sensitivity and sincerity, this sort of inquiry will not be seen as prying, especially when accompanied by sharing a few well-chosen words about your own background. The literature on self-disclosure by physicians, however, is not all that clear.

Two studies cast doubt on the value of self-disclosure by physicians. The puzzling finding in the first study was that a majority of primary care patients did not perceive physician self-disclosure as indicative of warmth or friendliness whereas a majority of surgical patients did (Beach et al. 2004). The second study recorded 117 visits with sham patients - with the doctor's permission but without identifying the sham patients in advance - and concluded that such self-disclosure was not helpful (McDaniel et al. 2007). This conclusion was based on the observation that four out of five times, when a doctor interjected personal information, he or she never returned to the topic being discussed by the patient. Furthermore, the doctor's comment was seldom in response to a subject raised by the patient.

These observations are revealing and valuable. But I believe the conclusion - that self-disclosure is unhelpful - is wrong. The key to helpful self-disclosure is that it should be generous rather than selfish. If the point of sharing personal information with a patient is to demonstrate genuine interest in what the patient has just said and openness to hear more, self-disclosure is likely to be valuable. If the purpose is merely to talk about subjects that are important to you, it will not be.

Jargon

In the course of a day's work you will have to switch gears between discussions with nurses, specialists and technicians on the one hand and discussions with your patients on the other. There is a tendency to let the same jargon that you employ with other professionals intrude on your talk with patients. Our professional speech is full of abbreviations, acronyms, anatomic and biologic terms, eponyms, generic and brand drug names, and medical device names. There is no assurance that any of this jargon, second nature in our communication with other professionals, has any meaning to a sick person.

It is imperative to develop an alternate vocabulary that translates jargon into terms that will be clearly understood and reasonably accurate without sounding patronizing or condescending. Given that you will be speaking to people with a wide variety of educational backgrounds and practical experience, this is a challenging task. It is a mistake to depend on the patient to let you know when your use of jargon renders your speech unintelligible. This is because many people find it embarrassing to tell their doctors that they do not understand what they are saying or because people think they understand when, in fact, they really do not.

There is a flip side to the principle of avoiding jargon and technical language. In an effort to make things clear some doctors inadvertently use language that is so simple as to sound patronizing or condescending. It is wise to have some knowledge of your patient's level of education and sophistication and tailor your own language accordingly. This is only one of the reasons for taking time to obtain a careful social history when you meet a patient for the first time.

Lost in Translation

A different problem arises when you must talk to someone when neither of you is fluent in the other's language. In this case, speaking slowly is especially important. So is repetition, perhaps expressing the same thought with one or more alternate phrases when the information is complex. It goes without saying that increasing the volume of your speech does nothing to increase comprehension. Unfortunately, the frequency with which I hear medical staff speaking in an inappropriately loud voice to non-English speakers suggests it should be said. When there is no common language between doctor and patient things become more complicated. Ideally, you will have access to a trained medical translator. In practice this is rarely the case, and an untrained layperson has to suffice.

Perhaps it is simply human nature that most volunteer translators feel compelled to edit what is being said into a version they think will be more comprehensible. In doing so, the translator's own perceptions, biases and misconceptions invariably color the words they use to quote you or the patient. A patient says, "My chest hurts." The translator, drawing his or her own conclusion, may translate, "He has a pain in his heart." The doctor says, "Do you have heartburn?" The translator, not finding an equivalent for heartburn, says, "Does your heart burn?" You ask, "Do you have pain in your abdomen?" The translator says, "Do you have pain in your stomach?"

Therefore, unless the translator is a professional, it is important to state firmly that your words and the patient's words must be translated verbatim without interpretation, tidying up, expanding or editing the original version. But this often happens in spite of your admonition. A sure sign of this is when a lengthy statement by the patient is followed by only a few words of translation, or when the translator engages in an exchange with the patient before rendering a translation of what was said to you. Then you must pause and repeat your request for verbatim translation. By all means avoid the use of metaphors and colloquialisms when talking to a non-English speaker. These rarely translate well and can lead to serious misunderstanding. The use of medical jargon will also be more damaging than usual when translation is required. The situation with the greatest potential for problems in translation occurs when a sick parent has to rely on a child to translate. Beyond the fact that the child may be too young to translate accurately it is obvious that there are many subjects that parents will not discuss or will edit in the presence of their child. When there is no alternative to relying on a child translator take everything that is said with a large grain of salt and make every effort to review what was said when a more suitable translator is available.

Cultural Diversity

Cultural diversity in the United States is now greater than ever before. The current generation of new physicians itself reflects this increasing diversity. Inevitably, many doctor-patient relationships will occur between people with strikingly different backgrounds. Medical educators are grappling with the question of how to structure medical school curricula to respond to this fact. Attempts to sensitize students to cultural differences by "briefing" them with a compendium of "typical" cultural characteristics, values and taboos risk degenerating into stereotyping at best and racism at worst. On the other hand, there is general agreement that there is significant potential for poor communication, perhaps leading to poor care, when there is a wide disparity in the cultural backgrounds of physician and patient. I believe that your approach to the challenge of caring for people whose backgrounds are very different from yours should be based on two principles, one dealing with attitude, the other with language.

The importance of attitude can be expressed in the concept of "cultural humility" (Tervalon and Murray-Garcia 1997). In essence, this concept distinguishes between one's factual knowledge of a patient's cultural, ethnic or racial background and one's awareness of and sensitivity to his or her own reaction to this background. Ideally, a doctor's attitude should embrace the firm conviction that people from other cultures, however unfamiliar, share his or her own desire to be treated as a unique individual rather than a representative of some group. Consider the fact that the range of differences between individuals within a group will always exceed the difference of average characteristics between groups.

It helps to cultivate open-mindedness and tolerance in ourselves if we are not already well endowed with these virtues. This means avoiding making judgments about a patient's intelligence, credibility, reliability or any other quality based on stereotypes and first impressions. You may certainly educate yourself about a specific culture's customs or etiquette. A proprietary product is available for $4 at Culturegrams.com, which offers a four-page summary of customs and conventions for each of at least 200 countries. Even with such information in hand, it is important to respect the principle that people deserve a chance to be approached as individuals.

The rule that should govern your use of language in a multicultural interaction is simply, Keep it simple. This means, in addition to being even more scrupulous than usual about not using jargon, avoiding the use of slang and figures of speech, especially metaphors. I will leave to your imagination how American slang could be misinterpreted. Violent metaphors or sports metaphors that make sense in American English - dodging a bullet, fighting an infection, killing cancer cells, toeing the line, stepping up to the plate, hitting a home run - may confuse foreign patients. A metaphor might be illuminating in circumstances that bring to mind a natural image, for example describing bronchial anatomy as similar to the branches of a tree. But another metaphor such as describing emphysema as honey-combing or polycystic kidney disease as similar to Swiss cheese could be baffling. When in doubt, draw a picture.

Companions

It will often be the case that a patient comes to see a doctor with one or more companions. You should make a considered decision about whether to invite a companion to be present when you speak to a patient. The first step is to politely inquire about the identity of the companion when this is not immediately clear. When companions are first-degree relatives I feel it is advisable to welcome them without question. If, in the course of the visit, you sense some tension or conflict between the patient and the family member make an effort to arrange a later opportunity to speak to the patient alone. You may gain information that allows you to modify a treatment plan or communicate in a more effective way.

When the companion is a patient's friend you must make a case-by-case judgment on whether to include the friend in the visit. But it is unwise to assume that the companion is there with the consent of the patient. Sometimes, a desire to be polite or a need to depend on a companion for transportation or some other material support will make it hard for a patient to exclude that person even when the patient wishes to.

When there are multiple companions it is important to decide whether to limit the number who may remain present. The more serious the illness, the more likely there will be several companions, sometimes a surprising number. This can create an awkward situation when anxiety is high, especially in large, close-knit families. One approach is to pick some reasonable number and then ask the patient to designate who should stay and who should not. To justify your request you can explain that it is important for the patient to have a chance to give you all the information needed without being preoccupied with concern about loved ones.

When a companion interrupts, contradicts or tries to speak for the patient it is important to pause and explain that it is most valuable for you to hear the patient talk about his illness in his own words. When the interruptions continue, be firm but soften the request by offering the companion a chance to contribute information after the patient finishes giving his own account.

Documentation

Everything that passes verbally between you and a patient has the potential to be important at some time in the future. No matter how good your memory is, it is not a substitute for a written record. A seemingly small detail may contribute to a diagnosis when other facts become available later. A careful chronology may be important to a future consultant. When another physician is covering for you he or she depends on your written reports to guide the care of the patient. When a misunderstanding occurs it is invaluable to have a record of what was said or done in the past. All this adds up to a requirement for good documentation. There are a number of systems for creating medical records and organizing progress notes. These are addressed in medical school curricula as well as a number of publications and proprietary products. Adopt one that works for you and use it.

It takes time to produce a high-quality medical record, but the result is so valuable that you neglect it at your peril. If you create your records by hand, the importance of legibility speaks for itself. I will comment on using a keyboard later.

There are some subtle points about the language in your records that deserve emphasis. When you are recording patients' statements about themselves or descriptions of events avoid the use of pejorative phrases such as "The patient *claims* to be taking his pill" or "The patient *admits* to smoking." Much better are neutral phrases such as "The patient *states, says, reports* or *remembers.*"

When a patient is describing important diagnostic results or diagnoses by quoting an outside source it is important to qualify this information by writing that the patient "*understood*" the information until you can verify the accuracy of the patient's understanding by reviewing the primary data. It is common for people to innocently misunderstand or misremember technical data. Simply recording the information as stated by the patient implies that you have endorsed its accuracy.

Avoid, at all costs, recording your own clever editorial comments, attempts at irony or humor, criticism of patient behavior or anything else that you might regret or have to defend when the record is reviewed by the patient or a third party at some time in the future. All patients or their authorized representatives now have a legal right to receive a copy of their medical records. Assume that your patient will at some time in the future read every entry you make. If you enter something in error never erase, insert superscript or subscript or write over. When necessary, line out the incorrect data and record the correction in the first open space with an explanatory statement if needed. The implications for medical liability of any appearance of a surreptitious alteration are obvious.

Another dimension to the subject of documentation relates to the current enthusiasm for applying to the practice of medicine certain techniques of performance evaluation that have been popular in industry and the business community for several decades. All these techniques share one common feature, a nearly exclusive reliance on performance indicators that can be easily subjected to objective measurement. In other words, numbers.

The practical result of this reliance on objective indicators is that if a medical decision, action or outcome is not documented in the medical record, it cannot be measured. Therefore, for the purpose of evaluation, it did not happen. If the explanation for two alternate courses of action is not recorded the explanation was not given. If a blood pressure is not recorded it was not taken. If an immunization is not recorded it was not administered. If a question about smoking was not documented it was not asked.

A physician could be practicing superb medicine, communicating effectively and revered by his or her patients and still receive an unsatisfactory rating if these decisions, actions and communications were not reflected somewhere in the medical record. At a time when numbers and documents rule, reputation and subjective impressions are no longer enough. It does not require much imagination to foresee a time when compensation and career advancement are tied more closely to documentation.

Checklists provide a convenient vehicle for documentation. Although I once scorned the use of checklists as beneath the dignity of a serious professional I have had to revise this opinion. There is increasing evidence that the use of checklists to ensure adherence to best practice in medicine can have a major positive effect on professional performance and, more important, medical outcomes. A discussion of this subject is beyond the scope of this book, but I strongly recommend the work of Atul Gawande, M.D., for a thorough analysis of the use of checklists in medicine (Gawande 2007, 2009).

The Physical Examination

The process of performing a physical examination might not be expected to be an occasion for much communication between doctor and patient. In fact, the physical examination presents an opportunity for some important communication, both spoken and unspoken. We should always be aware of the implicit measure of trust required when a patient permits the invasion of privacy required for a physician to perform an examination. All the normal boundaries of age, gender and social status are temporarily suspended when the doctor looks at, listens to, or touches and probes the uncovered body of another person. Beyond this, consider the traditional, symbolic importance, common to nearly all cultures, of the laying on of hands by a healer.

Early in my career a respected mentor, Dr. Samuel Asper, showed me a simple gesture that acknowledges the profound significance of these considerations without saying a word. This is the practice of beginning every examination by gently measuring a patient's pulse by palpating the radial artery at the right wrist. Try doing this in silence and counting the pulse for a full thirty seconds. This simple gesture, if done carefully before peering into a patient's mouth, drawing back a gown or drape to expose the chest or abdomen, grasping a limb or some other intrusive action, will convey respect, concern, and a thoughtful attitude on the part of the doctor. When you see anxious patients become less so and tense patients relax in response to this gesture, you will be willing to spend the half minute or so required for this old-fashioned maneuver. In the presence of cardiac disease you might even make an important discovery without the use of technology.

It is also a good practice to announce what you are about to do as you proceed with an examination; "I am going to look at you eardrum" or "I am going to listen to your heart" or "Try to feel your liver." When an action is required of the patient - take a deep breath, open your mouth, lift your arm - throw in the word *please* now and then. For good measure, recognize the patient's effort by saying "That's fine" or "Thank you" from time to time. Rather than seem excessively formal, these polite expressions are a way to convey respect with essentially no cost in time or effort.

With the exception of saying "That sounds good" or some other reassuring phrase, it is best to defer discussion of your findings until an examination is complete. I will say more about how to communicate abnormal findings in chapter 4. Finally, save small talk for another time. It is tempting to use humor or chatter as a way to relieve stress during an examination. But the cost of interfering with your own concentration while trying to interpret and remember physical findings is too great to justify the distraction. By the same token, trying to answer a patient's questions during an examination will surely lead to a less effective response or a less skilled examination or both. And when it is the patient who is making use of small talk, a gentle request for silence is called for. If you must, put your stethoscope in your ears and hold a finger to your lips.

Time Flies

Both physicians and patients almost uniformly hold the belief that time available for medical visits is inadequate. But the one published study of the length of office visits in the United States found that between 1989 and 1998, a period when enrollment in managed care programs was growing, average duration of office visits actually grew by 12 percent (Mechanic et al. 2001). This conclusion was based on data from the

National Ambulatory Medical Care Survey (NAMCS). The use of NAMCS data might be criticized because the absolute increase in visit duration was only two minutes and the data was based on self-reporting by medical offices with a response rate that varied from only 68 percent to 74 percent over the reporting period.

The medical offices reported the number of hours spent with patients per week and the numbers of patients seen per week. Dividing hours spent with patients by number of patients seen gave an estimate of minutes per visit. The NAMCS follow-up study in 2005 found no change in overall mean time spent with a physician since 1995, as well as a 20% increase in visits lasting 16-30 minutes (Cherry et al. 2007).

Although this data concerning length of visits may be inconclusive there is little debate about the value patients attach to time spent with their doctors. And it is clear that time spent is an important factor in determining patient satisfaction (Lin et al. 2001). If the NAMCS data is accurate (I am skeptical of its validity), what could account for the disparity between the data and the view of a majority of patients who do not feel their doctors spend enough time with them? I think the answer lies in the importance of perception over reality in shaping a patient's assessment of his or her visit with a physician.

If the doctor is standing, looking at a computer screen, subject to interruptions or giving nonverbal signals of impatience, it is easy to imagine that a sick person would not only fail to appreciate a few more minutes of "face time" recorded in these studies but also sincerely believe that the visit was shorter than it actually was. In practice, busy physicians may or may not have the option of spending more time with each patient. But they do have the option of sitting down, speaking slowly, restricting interruptions and avoiding behaviors that could communicate impatience. If time permits, a few moments devoted to non-medical small talk will favorably influence perception. These are all modest measures that can enhance the value of every minute the doctor actually spends with a patient.

***

The remainder of this book is devoted to what doctors should say and how to say it. This emphasis should not be taken to obscure what may be the most essential quality that should guide the process of talking to the sick, namely, listening. Dr. Alicia Conill, Associate Professor of Medicine at The University of Pennsylvania School of Medicine, read a moving essay on the subject of listening to patients in the *This I Believe* series on National Public Radio recently (Conill 2009). Having to give up teaching after becoming a patient herself, she concludes, "I believe in the power of listening. I tell them [her students] I know firsthand that immeasurable healing takes place within me when someone stops, sits down and listens to my story." Her essay should be required reading for physicians.

CHAPTER TWO   THE MEDICAL HISTORY

The initial medical history is among the most important exchanges a doctor and patient will have. It creates a first impression, lays the factual basis for medical decisions and sets the tone for the personal relationship with a new patient.

The Patient Centered Interview

In taking an initial medical history you have the opportunity to employ one strategy that has more influence on the success or failure of this encounter than any other factor. The strategy consists of what I refer to as "letting it run." After a greeting and introducing yourself you say something like, "What brings you to the office?" When your patient responds, "The Sixth Avenue bus," you know you are dealing with someone who thinks in very concrete terms. Perhaps "How can I help you today?" is better. When a patient presents with a problem only after living with it for a lengthy period, asking, "What led you to come in at this particular time?" often yields an important insight. The symptom that turned out to be "the last straw" may be an important diagnostic clue. The availability of funding or transportation or the prompting of a spouse can give important information about social issues.

In any case, after your initial gambit, sit back and listen. Doctors interrupt their patients with remarkable frequency (Rhoades et al. 2001). If you can listen without interrupting, signaling that you are in a hurry by shifting in your chair or glancing at your watch, commenting or otherwise interfering with the patient's flow of words, you have a good chance of hearing the essence of what the patient feels is wrong and what he or she thinks is important. Of course, this information eventually will have to be organized, clarified and recorded in a patient's chart as *The Present Illness*. First, however, it is only fair to listen to the patient's unedited version. Unfortunately, there is no substitute for spending precious time if this is to happen. A modified version of this strategy can be valuable even in acute settings short of an overt emergency.

You will rarely hear a coherent, chronological account of symptoms and events in clear and unambiguous language. You will usually hear a less organized, sometimes confusing mélange of symptoms, third party information, unrelated events and personal asides - given in the order they are recalled, not the order in which they occurred. Occasionally a patient will begin talking in a stream of consciousness fashion and give no sign of slowing down after periods of agonizing (for you) length. Your task is to continue to listen quietly as long as you can possibly afford. You may nod or say "Yes" or "I see," but try to repress your growing desire to interject questions, steer the account back to the point, request clarification or simply flee.

For anything less than a trivial problem, especially in an internal medicine or primary care setting, if you have only scheduled twenty minutes for a new patient visit you have defeated yourself before you even begin. In fact, because you cannot know if a problem is truly trivial until you see the patient, it is better to schedule initial visits for as long as you possibly can in all cases. If you finish early the rest of your schedule will appreciate the extra time. In my practice, if time ran out when I was confronted with an unusually complex but long-standing problem, I found it useful to concentrate on the present illness and ask the patient if we could schedule a second visit to complete the rest of the history and physical examination.

Remaining Neutral

While you are "letting it run" keep your reactions neutral. Sick people sometimes say outrageous or alarming things. When you exhibit surprise, approval, disapproval, or worst of all impatience, you will strongly influence what the patient says from that point on. People unconsciously monitor the response of their listener and sometimes adjust their speech according to that response. Sick people or worried well people are all the more sensitive to the impression they are making on the doctor who will have a role in making them well or preventing illness.

In this respect the medical history is analogous to courtroom testimony. Both processes seek the truth or something as close to the truth as the complexity of human expression permits. The physician's role is to avoid putting words in the patient's mouth, in effect, "leading the witness." Also to be avoided are nonverbal reactions that cause the patient to modify or suppress information that might have been more helpful if given spontaneously.

The Open-Ended Question

Assuming that you have allotted enough time to take a good history, you will, at some point, begin the process of clarifying, expanding and organizing what has been said. While the patient was speaking perhaps you discreetly made a few notes to guide the process, which will consist of asking a series of questions. These questions should be as open-ended as possible; "What else?" or "Then what happened?" instead of "Did you do _____?" "What did you feel then?" instead of "Did your arm hurt?" (Barrier et al. 2003). The objective is to give your patient latitude to recall small details and crucial particulars that will surely be suppressed if you are mechanically following a checklist of pertinent questions. Some patients, instead of running on endlessly, will be taciturn. It is almost as if they expect you to intuit what is wrong. Such a patient will imply or even say explicitly, "You are the doctor. Tell me what you want to know." Remaining silent or providing a neutral prompt such as, "Please go on" or "I see" is much better than breaking in with a pointed question. For a patient who is clearly distressed about something but unable to express what it is, a confirming statement from the doctor such as "That sounds really awful" or "You must have been very upset" will often open the door to more information.

Again, interrupting with a specific question at this point risks missing useful information that might have been forthcoming but does not happen to correspond to that particular question. Another reason not to fall back on pointed questions early in an interview is that they will, quite unconsciously, be influenced by your judgment about what the probable nature of the patient's illness is. The problem here is that when a physician makes a leap to a premature conclusion, usually based on recognizing a familiar pattern in the patient's presentation, it can be quite wrong. There is a place for checklists and relevant questions later in the visit in what is usually called the *Review of Systems*. If there is no time left after a lengthy present illness you may have to reserve this portion and the rest of an initial history for another visit, although this too risks missing pertinent data.

Note Taking

When a physician's attention is focused on a pad of paper or a chart, or even worse, a computer screen, it cannot be focused on the patient. Furthermore, while you are concentrating on what you are writing or typing you may well miss useful clues in your patient's demeanor or small details or nuances in what he or she is saying. If you are not capable of typing without looking at the keyboard, trying to produce a finished record using a computer during an interview is a good way to give the impression that the patient is not your first priority.

Special requirements apply to new patients or visits for significant new problems with established patients. In these cases it is best to keep notes sketchy and not attempt to compose a smooth record during the visit. A finished record of the interview should be created shortly after the visit when can do it properly. Further, because you can then pay full attention to what you are writing or dictating, this practice will contribute to the quality of your record. The principle here is that a medical history that records a visit for a significant new problem is fundamentally different from a follow-up visit when notes can usually be recorded directly in the medical record.

Obviously much of the advice offered in this chapter cannot be fully applied in an emergency room, certain surgical consultations or a walk-in acute care clinic. Even in these special settings, however, patients will invariably appreciate a chance to tell their stories the first time around with as little interruption as possible. And you may learn something useful.

***

Specific Symptoms

Several symptoms require very directed inquiries in order to gain a full understanding of a sick person's condition. As the list of possible diagnoses consistent with a patient's symptoms becomes more focused, it will be necessary to fill in additional details - timing, exposure to environment, relation to external events - that were not part of the patient's spontaneous description. When doing this, avoid compound questions that combine more than one inquiry in a single question; for example, "How long did it last and what were you doing when it started?" Asking leading questions here is as undesirable as it is for the *Present Illness*. Ask "What makes it better?" rather than "Does it get better when you lie down?" Ask "Does the problem follow a pattern?" rather than "Does it only happen at night?" Below I discuss directed inquiries into some common symptoms.

Pain

Few other symptoms are characteristic of so many diagnoses. And few symptoms vary so much in expression from one person to the next. Three qualities associated with pain deserve attention: location, character and intensity. In respect to the location of pain I have found that asking a patient to point to the place where it hurts is better than asking, "Where does it hurt?" A surprising number of people have no idea or a mistaken idea about where their internal organs are located.

They will say it hurts in "my stomach" when they mean the lower abdomen or "my back" when they mean the flank. Even fewer will know that cardiac pain can radiate to the jaw or left arm, kidney pain to the flank or gall bladder pain to the back.

Because people often hold mistaken ideas about the location of their internal organs, it is not prudent to accept any statement that pain originates in a specific anatomic site. Always ask the patient to demonstration the location on his or her body. When a patients have trouble pointing to the location of their pain you can provide a simple outline drawing of a human figure and ask them to mark the primary location. It is also possible to use this device to accurately locate the direction of radiating pain. The technique is so useful that I had two rubber stamps made with anterior and posterior outlines of human figures and incorporated the completed figures in patient's charts for future reference.

An endless variety of language can be used to describe the character of pain. Textbooks of medical diagnosis provide lists of the classic descriptions of pain associated with specific conditions: squeezing or crushing for cardiac pain, aching for joint pain, lancinating for neuralgia, cramping for intestinal obstruction and so on. Suffice it to say, the variety is so great and so subjective that the reliability of these associations is limited at best. Still, a patient's perception should be elicited and recorded, preferably without too much prompting with specific terms and in his or her own words.

Assessing the intensity of pain is even more complicated. In recent years there has been remarkably unquestioned acceptance of a particular technique for asking this question. The technique is based on asking a patient to rate pain on a scale of 1 to 10, where 1 represents little or no pain and 10 represents the worst pain ever. It is claimed that this practice yields more accurate information than allowing the patient to express a subjective description in his or her own words. It may well be that using this technique serially in a given patient over time provides more reliable information about the trend of the pain level.

However, many perfectly sincere people tend to minimize or exaggerate their discomfort because of cultural influences, life history or personal circumstances. It is not clear that using a numerical scale is any better than subjective language for providing insight into the severity of pain in these individuals. In saying this, I do not wish to imply disrespect for advocates of the principle that a person's pain is whatever they say it is. This is a valid principle, and adhering to it is a way to prevent our own bias from clouding good judgment in treating pain.

For inexperienced medical personnel, asking a patient to use a numerical scale to express pain may be superior to using the inexperienced observer's assessment of the patient's language, behavior and physical correlates of pain to evaluate intensity. But, compared to the use of careful listening, uncritical reliance on a numerical scale is at least as likely to result in under-treatment of pain experienced by a stoic person or over-treatment of pain for a very frightened person. Especially in treating chronic pain, there is no good substitute for thoughtful analysis based on enlightened inquiry into a patient's emotional state and medical history.

Vices

Seeking information about a patient's vices will raise some sensitive issues. No one likes to admit that they smoke, drink too much, take illicit drugs or indulge in any number of questionable activities. More important, most people are aware that, in the medical setting, such disclosures may become part of a permanent record. The potential for future impact on insurability, employment and family relations is obvious. When any such potential exists it is appropriate to explicitly state that you hold the information in confidence and the fact that doctor-patient communication is privileged. If there is no compelling reason to include the information in the patient's chart decide whether to advise the patient that it will not be recorded.

When it is pertinent to the investigation of illness to ask about one of these practices your tone and manner must be free of any hint of approbation or prejudice. Resist the impulse to be judgmental. Giving the impression that you disapprove of the practice or expressing disgust will foreclose any possibility of getting a full picture of your patient's situation, let alone helping him or her. When talking to patients with well-established alcoholism, simply asking how much they drink is a waste of time. If it is essential to know exactly how much they drink it may be informative to start by asking what their favorite beverage is and the details of how they purchase their beverage; size of container, how often they make a purchase, whether someone in the family facilitates the purchase, and so on. A similar approach can help to quantify smoking accurately.

Someday you may find yourself caring for a patient who either hints or explicitly says that he or she has been involved in criminal activities. Before you allow a patient to elaborate on statements of this kind be quite sure that the information is pertinent to treating the illness. Examine your reaction to what is being said, and be certain that you are not being drawn in to a discussion simply because the lurid details satisfy your curiosity or are a source of fascination or titillation. Being taken into the confidence of a patient involved in illegal activity can lead to all sorts of complexities that you may be unprepared to deal with.

Sexual Activity

Seriously ill people, because the disease either dominates their thoughts or suppresses their libido, seldom raise sexual issues with their doctors. But this does not mean that problems with sexual activity do not occur during illness. Even if it is not perceived as a problem by the patient, loss or threatened loss of sexual activity is a problem for the patient's partner. When one considers the unease many physicians experience in asking about sex it would not be surprising to find that sex-related problems often go unrecognized and untreated. Including at least a simple question about problems with sexual performance in the review of systems for every new patient is the first step in addressing this situation. Equally important is being alert to verbal clues that a problem exists. A casual or joking reference to sex by patient or partner should be interpreted as an invitation to talk about sex on a more serious level.

Expert advice for physicians on the subject of taking a sexual history usually emphasizes the importance of showing empathy while still maintaining a matter-of-fact demeanor and not reacting to what a patient says. It also emphasizes the importance of inquiring about sexual problems even when the issue has not been raised by the patient. I think this advice ignores two important problems. First, there is what might be called an empathy gap. Inevitably, in about 50 percent of cases, the doctor will not be the same gender as his or her patient.

Further, many sexual problems will center on practices or ideas that are remote from the experience of the doctor regardless of his or her gender. In these circumstances being empathetic is a tall order. Second, especially when dealing with seriously ill people, barging ahead with inquiries about sex early in the evaluation can produce alarm and suspicion in patients for whom sex is the furthest subject from their mind. I would counsel a cautious approach, delaying exploration of possible problems with sex until the relationship with your patient has matured to a point where there is a solid basis of mutual trust and understanding.

Quantitative and Qualitative Analysis

It is important to have strategies for obtaining objective data that leads to accurately quantifying certain symptoms. Do not allow a report of weakness to go by without distinguishing between loss of muscle power, nonspecific fatigue or reduced exercise tolerance. If muscle power is the problem, what can a patient lift? Can the patient get up from a chair without using his arms? Is the weakness proximal, distal or generalized? These details not only quantify the weakness but also may give clues to loss of power in specific muscle groups that points to a diagnosis. If the problem is reduced exercise tolerance, how far can a patient walk or how many stairs can he climb? Is muscle fatigue or shortness of breath the limiting factor? Breathlessness should trigger similar inquiries about what posture, effort or exertion provokes loss of breath as well as cough and wheezing. Generalized weakness or fatigue, not associated with exertion, may signal depression rather than a physical problem.

All the various discharges that accompany illness need to be quantified in terms a layman can grasp: for bleeding, staining on tissue, a few drops, a cupful, and so on; for sputum production, a similar range of concrete measures. Diarrhea should not be recorded without amplifying data that nails down frequency, volume and character. A full list of how to pursue every possible complaint is beyond the scope of this book, but the principle to follow is that recording a symptom without quantitative data does not support a well-informed diagnostic analysis.

Chronology

The relation of symptoms to time is a good example of the value of a systematic approach to asking specific questions. People often find it difficult to recall when a problem started or when it last occurred. A patient may be able to pinpoint the onset of a symptom if prompted by a well-defined event such as a holiday - "Were you having this problem before Thanksgiving?" - or a personal milestone - "Did that happen before you graduated?"

Ask about timing patterns that might have significance - nocturnal pain in peptic ulcer, seasonal occurrence in allergic respiratory conditions for example. The characteristic episodic pattern of fever in malaria, Hogkins lymphoma, military tuberculosis and other diseases was recognized long before the advent of modern laboratory techniques.

Simple questions, such as "How often does it happen?" or "How long does it last?" often tax a patient's ability to give reliable information. Any patient with episodic symptoms of uncertain origin should be asked to keep a diary in order to provide an accurate record.

Presenting a History

It is frequently necessary to present a patient's history to a third party - when signing out to a colleague, making a referral to a consultant or presenting the patient to a faculty member. When this is done in the presence of a patient some special practices should be observed. Unless the third party is already well known to the patient they should be properly introduced to each other, using the patient's full name and the third party's title. Then a short explanation for the purpose of the presentation should be given, along with a request for permission to go ahead. Referring to a patient with an impersonal pronoun (he, she, him, her) tends to objectify the patient and should be avoided - this is equally so whenever you are talking about a patient to someone else in the patient's presence.

If sensitive or embarrassing information is to be disclosed, a brief aside to the patient referring to the information and why it is important and another request to go ahead is necessary. For a presentation in a teaching environment it is very considerate to ask the patient if he or she wishes to add something at its conclusion and, of course, to thank the patient for cooperating.

CHAPTER THREE ANSWERING QUESTIONS

Illness provokes questions. Some well-functioning aspect of our physical or mental state, previously taken for granted, changes or stops working altogether. This is troubling, and we want an explanation as well as relief: What is wrong? What caused it? How serious is it? Will it get better? How do I make it better? The observation that knowledge is power applies to the sick as well as to their doctors. What a doctor says by way of explanation and answering questions forms the basis of a patient's understanding of his illness. How skillfully the doctor does this will play a large role in determining how well a plan of care works independently of the scientific quality of the plan.

What's Wrong?

Some symptoms and signs seem so obvious and familiar that we are led to draw our own conclusions about what is wrong. After reaching this do-it-yourself diagnosis we present to our doctor with a firm conviction about the nature of our illness: "I came in contact with poison ivy," "I swallowed a fish bone,"

"I have been constipated." If we have, in fact, reached an accurate conclusion the doctor's role can consist of confirming the diagnosis and moving on to the next subject. Problems arise when, as often happens, patients' conclusions, however confident, are mistaken because their data is incomplete or they have misinterpreted the evidence.

A prudent physician has the task of beginning this encounter by reserving judgment or disagreeing with the patient. Do this cautiously. To challenge firmly held convictions abruptly or scornfully may provoke all kinds of negative reactions. Many people, especially those blessed with good health, take considerable pride in their ability to stay well and manage their own care. How often do you hear a patient say something like "This always happens when I eat strawberries," "A cold always goes into my chest," or "I know my own body"? In fact, there is an element of truth in such statements. Living in our own skin throughout life we are, indeed, more familiar with our bodies than any other person can be. On the other hand, a patient presumably has less expert knowledge of medicine than a doctor does. Most important, all of us lack objectivity where our own health is concerned.

After analyzing what you have been told and finding a patient's conclusions questionable, do not be smug. Rather than dismiss the patient's conclusion outright or, worse, express amusement or superiority when the self-diagnosis is far off the mark, spend a few words validating the patient's ideas. Say something like, "You may be right, but let's do another test to be a bit more certain just in case this is an unusual situation" or "I know it seems just like what you had before, but sometimes another diagnosis can imitate that condition." Such phrases, in addition to avoiding hurt feelings, serve to protect you from the common diagnostic error of leaping to a premature diagnosis.

Brimming with the fruits of the latest medical research and proud of our own fund of knowledge, it is sometimes difficult to resist saying too much too soon. Until you are certain of what is going on, it is sufficient to explain that the signs and symptoms suggest malfunction in a particular organ or system. It is fine to add some detail concerning the physiology of that organ or system when it will help to allay anxiety, foster better compliance or just satisfy a patient's curiosity. This kind of information may give a sick person confidence and support the feeling that he or she has a measure of control over the situation.

But, even when you have some concern about a serious diagnosis, it is a mistake to list every rare and ominous-sounding possibility, potential complication or bad outcome at an early stage of management. In other words, if a question is not asked you need not explain that some sore throats are a prelude to pharyngeal abscess or that occasional numbness in the hand is an early symptom of multiple sclerosis.

When probing "what if" questions are asked it is a mistake to ignore them or dismiss them as too remote to consider. Cultivate the ability to deal with questions about dire long shots in a way that exhibits legitimate concern without instilling unrealistic fears. You can accomplish this, not by declaring, "That's out of the question," but by acknowledging that, while such cases do exist, you can give firm reassurance that none of the present findings point to such a case. It also helps to promise that you will watch carefully for evidence to the contrary.

The goal is to confront the reality of a serious diagnosis squarely while providing perspective about its rarity accompanied by reassurance. Demonstrating that you recognize the possibility of a bad diagnosis also demonstrates respect for the patient's concerns. This is another example of conditioning your own thinking by what you say to a patient. Thoughtful physicians maintain a balanced perspective by keeping the possibility of a dire long shot diagnosis in mind rather than suppressing it by habitual denial.

When I was a medical student great emphasis was placed on approaching diagnostic problems using the exercise of differential diagnosis. The concept was to acquire a mental inventory of the typical findings associated with every diagnosis and use a process of elimination to exclude those conditions that are inconsistent with a patient's set of findings. Then, having narrowed the possible diagnoses to a manageable few, with a few well-chosen additional tests or examinations, the doctor could settle on the single diagnosis that remains after every other one

48

has been eliminated. When done well the process was elegantly logical and a great way to show off one's intellectual accomplishments.

Now that it is possible to create detailed images of an individual's entire anatomy with high-tech scans and probe body chemistry with highly sophisticated laboratory tests the exercise of differential diagnosis, once a highly revered process, is a less respected part of the medical curriculum. Computer algorithms have been devised that rival the diagnostic ability of the most experienced clinicians. In an earlier era clinicians with a vast, easily recalled mental inventory of diagnoses and great talent for synthesizing a correct diagnosis from a confusing collection of clinical data were admired above all others. Today sick people and doctors alike are impatient to reach a diagnosis. The temptation to reach an early diagnosis by the brute force application of every available test is difficult to resist.

But there are problems inherent in doing too many tests. In addition to the obvious economic cost, poorly planned testing sometimes leads to ambiguous results. This problem is especially relevant to the task of explaining what is wrong. It is difficult enough to translate definitive test results into lay language. Although they are powerful tools, blood tests, imaging techniques and immunopathology and genetic analysis often yield results that, while outside the range of normal, are of uncertain significance. Explaining this sort of result is even more difficult.

A small shadow on an x-ray is probably an old scar but might be something sinister, a slight elevation in a blood test seems completely unrelated to the rest of the picture but might indicate a new problem, a scan done to investigate one organ shows an unexpected abnormality in another organ. Perhaps the result is in the gray zone between normal and abnormal. Or there is controversy about what normal is in the first place. Not uncommonly a confusing abnormal result is simply a laboratory error.

To make matters worse, some useful diagnostic procedures are accompanied by an unavoidable small percentage of false positives, abnormalities that are confirmed as real but occur in the absence of disease. Faced with an unexpected or ambiguous result or a false positive, the doctor's choice for responding lies between pursuing a chain of increasingly complex, expensive and possibly risky additional procedures or simply waiting to see if the passage of time with repetition of the same test after some interval resolves the ambiguity. Especially with procedures that carry a risk of injury, the cost of pursuing a diagnosis can be higher than the risk of delay.

While choosing a wait-and-see strategy may be medically correct, once an "abnormal" result has been reported, it may be difficult to persuade a worried patient that this choice is an acceptable alternative. An effective way to talk about this situation is to explain a recommendation to wait-and-see in terms that make a convincing case for the following analysis.

First lay out the details that render the result ambiguous. Then describe what would have to be done to resolve the ambiguity. Next explain that there is a balance between the risk, inconvenience and cost of immediately pursuing additional tests on the one hand and the risk that any harm might result from potential delay in resolving the issue on the other. When that balance clearly favors delay, the correct course is to defer further tests. This sort of risk analysis will be seen as cold-blooded or uncaring by some patients and your analysis may fail to be convincing.

But it is always a worthwhile exercise for the physician and should be carried out each time such a decision is needed. Unless you find your own argument for delay convincing, perhaps it would be wiser to go ahead with more testing. Every sick person deserves this analysis. When the reasoning behind the recommendation is valid it is a powerful argument for the conservative approach. Whatever the patient's reaction, sharing your well-considered analysis with your patient is always preferable to sweeping an ambiguous result under the proverbial rug.

What Caused It?

Disease can still be classified according to the system in use when differential diagnosis was a premier skill. In this system categorizing diagnoses as the result of inflammation, trauma, inherited defect, metabolic abnormality or neoplasm helped to ensure that all possible diagnoses were at least initially considered. What is different today from the medicine of fifty years ago is that the list of diagnoses for which a specific causative agent is known, be it a virus, environmental toxin or genetic abnormality, has greatly expanded. Today's patients expect that the cause of their symptoms can be precisely identified. This sets the stage for considerable frustration when the cause remains obscure in spite of extensive efforts to identify it. It is important to remember that it is only human nature to seek concrete explanations for momentous events. For most of us the idea that something important just happened for no reason is unsettling and foreign.

Whereas doctors place a high value on the scientific method and logical deduction in diagnosing illness, the sick person's efforts to explain his or her condition may follow patterns of thought that are less than helpful. To some extent all of us will slip into one of these patterns in times of stress. Two are worth mentioning: *post hoc ergo propter hoc* reasoning and *confirmation bias*. Robert Todd Carroll has published a fine guide to these and other logical fallacies that are pertinent to medical practice in his book, *The Skeptic's Dictionary* (Carroll 2003).

In brief, post hoc reasoning leads to a belief that simply because one event happens after another event, the first event was the cause of the second. Thus, consuming sugar causes hyperactivity in children, sitting in a draft causes upper respiratory infections, contact with this or that agent causes warts, immunization causes autism and so on. Confirmation bias refers to selective thinking in which one tends to seek and take note of facts that confirm one's previous beliefs and to not seek or to discount the relevance of facts that contradict these beliefs. A sick person may have become convinced by the pervasive influence of popular media, advertising or advocacy groups that a particular nutritional deficiency or outside agent caused them to be sick.

The effect of these patterns of thought can cause sick people to have firmly held but incorrect notions about the cause of their symptoms. They can be so convincing in reporting their belief that a physician may be deterred from looking further. If a physician is not to be misled he or she must keep an open mind when analyzing patients' conclusions about the cause of their condition. Challenging such beliefs is sometimes important in reaching a correct diagnosis or gaining a patient's cooperation in treatment. However, the fact that a patient may have invested a great deal of emotional energy in reaching his or her belief means that correcting it must be done with tact and awareness of how strongly the belief is held.

The influence of post hoc reasoning and confirmation bias is nowhere more clearly shown than in one type of epidemiologic research. Studies of this type are based on retrospective surveys that aim to identify factors that cause a particular disease. In such surveys people with the diagnosis in question are asked about past experience, dietary habits or exposure to some environmental agent or drug prior to the onset of their illness. Occasionally the association between one of these influences and the diagnosis is found to be more frequent in the sick group than in an otherwise similar control group of people without the illness. When this happens a conclusion may be reached that there is a cause-and-effect relationship between the environmental agent, drug or other factor and the illness. The flaw in these *retrospective* studies arises from the fact that sick people usually will have already conducted a detailed review of their past experience looking for an explanation for their diagnosis.

For example, they will identify dietary preferences and recall exposures to some toxin significantly more often than would a population of people who do not have the disease in question. Subjects in the control group without the illness will usually not have conducted such a review and will have forgotten or never noticed the experiences identified by subjects who are sick. The survey may then reveal an association that, although statistically significant, was the result of post hoc reasoning or confirmation bias rather than a genuine cause-and-effect relationship. Only a study that tracks healthy people over long periods, *prospectively* recording their experiences and exposures before they become ill, as well as their incidence of disease, can conclusively establish cause-and-effect relationships of this kind.

All too commonly, research that contradicts the findings of earlier studies leaves the public confused and skeptical about the value of epidemiologic findings in general. When a worried patient comes to the office concerned about the latest report of an association between some behavior or exposure and a disease it will be useful to take the time to explain all this rather than simply dismiss the report.

Conversely, there are some causes for disease that, even when clearly proven, need not be belabored when talking to sick people. I see no purpose in reminding the heavy smoker that the lung cancer afflicting him is the result of smoking, the diabetic that obesity caused her diabetes or the alcoholic that heavy drinking harmed his liver. Harping on the causal relationship between past behaviors and illness is not helpful. These relationships are probably known all too well by the affected patients. It serves no good purpose to instill or reinforce guilty feelings by reminding them that their own actions or inactions contributed to being sick.

A better strategy is to emphasize those constructive measures that are still within patients' ability to take that can improve their situation. In fact, if you sense that a patient is burdened by guilt about previous behavior that caused his or her condition, it can be highly therapeutic to say something supportive, for example, "Your situation in life gave you limited choices," "How could you have known then what the future might bring," or "Many other good people have made similar mistakes."

How Bad Is It?

There are both practical and emotional considerations to this question, some trivial, some profound. The question may be couched in many different terms. "Will I be well in time to keep my travel plans?" "Will this affect my golf game?" "Will I live to see my children grow up?" Your answers must be carefully calibrated to meet the true meaning of the question. For questions of obvious practical (business transactions, professional commitments, estate planning) rather than emotional significance, it is sufficient to stick to the facts. But often the question is, at heart, an expression of a need for reassurance.

People differ widely in their need for reassurance. At one extreme a macho male or a woman who sees herself as the pillar of strength in her family will habitually deny that any serious consequence might follow illness. At the other extreme a severely anxious individual will automatically assume the worst, perceiving even a mild illness as a mortal threat. How to strike a balance between too much and too little reassurance is not always obvious. The importance of having some knowledge of your patient's background, life experience and personality is most evident when you need to choose language to use in responding to the needs of a particular individual. The doctor who automatically reassures every patient in the same way will reinforce denial in some and provoke disbelief or give insufficient support for others.

I think it is unfortunate that, in an era when malpractice litigation is common, many doctors find it difficult to give an appropriate level of reassurance. They fear that, in the event of anything less than an ideal outcome, reassurance will come back to haunt them. Remember that reassurance is not the same as an explicit or even an implied guarantee of a good result. In surgical settings it is possible to give an honest picture of what can go wrong while balancing that information with a description of the concrete measures that will be used to prevent complications.

As long as optimism is not wildly unrealistic or based on deliberate fraud, sick people welcome a chance to hope for a good outcome. Nothing could illustrate this more clearly than a moving essay by the late Alice Trillin (Trillin 2001). In this essay she tells of her reaction to an optimistic view expressed by one doctor after several other experts had given the worst possible interpretation of some x-ray results. When she questions the optimistic interpretation by asking, "What if it is cancer?" the doctor gives what I consider an ideal reply, "If it is, you will come back to me and I will tell you what we can do. I will also tell you what I think you should do, and then you can make up your mind." Instead of suspicion, Alice Trillin felt, "That was when I knew I would trust this man."

Will It Get Better?

In essence, answering this question leads to making some statement concerning the medical concept of prognosis. Doctors often rely on their own experience to formulate a prognosis. This can be entirely appropriate when they have practical experience with a new treatment that improves the picture for a specific diagnosis from what was the case in years past. When basing a statement about prognosis on personal experience it is both wise and ethical to make it clear that this is the case. But more commonly we have to rely on statistics to make a confident statement about prognosis.

As is the case with risk, these statistics depend on the published, aggregate experience of large numbers of patients treated at a variety of facilities, often analyzed in considerable detail. A statement based on such data will have the weight of objectivity and can be highly accurate. On the other hand, it is small comfort to individuals who find themselves in the 10 percent who do not get well that their condition is 90 percent curable. The outcome for a single individual is usually all or none. The individual is either all alive or all dead, their limb or sight is either spared or not spared, his or her cancer is either cured or not cured.

When speaking to an individual about his or her own illness, statistics should be quoted with discretion and a large measure of sensitivity, or perhaps not at all unless specifically requested. For physicians, trained to value objective data and the scientific method, quoting statistics without careful support and elaboration can substitute for thoughtful exploration of how the patient sitting in front of them actually feels about his condition, and how well prepared he is to deal with statistics. Therefore, remember that an individual is not a statistic. When the numbers are unfavorable and you cannot avoid citing statistics, try to emphasize facts unique to your patients that may improve their own individual odds.

As is the case with reassurance, many physicians are reluctant to express optimism when discussing prognosis with sick people. The reluctance is based on fear that, if things do not go well, the physician will be subject to recriminations or blame because that optimism was taken as a promise of success. In fact, I believe the opposite is true. In the setting of a grave medical diagnosis it is possible to give an honest picture of the seriousness of the condition while emphasizing the array of current treatment alternatives and the prospect for future medical progress.

Failure to provide some measure of reassurance, however cautious it might be, leaves a sick person without a useful and much-desired defense mechanism. I believe that doctors should treat optimism as a precious gift. Even if all you can find to say is a reminder that sometimes it is possible to beat the odds, do not fail to be optimistic. When it is backed up by a doctor's integrity and sincere concern a sick patient will treasure an expression of optimism.

How Do I Make It Better?

At some point you will be expected to answer this question. When there is a single, widely accepted approach to treatment this is relatively easy. It will be sufficient to make a recommendation that is clear, precise and well documented. There will, however, often be situations in which there is controversy about what is best practice or where the balance between unwanted side effects and the potential benefit of treatment is narrow. There will be other situations in which there is no satisfactory standard treatment and the option of experimental treatment arises. In these cases there is a special obligation to present alternatives in a way that is not only honest and fair but also leaves the patient with a clear idea of what you think is best.

"Best" is probably not the correct word here. Because different people will have different goals and values, even as a physician you may not be in a position to know what is best for someone else. Although statistical data from the medical literature can point to the more effective or safe of several alternatives, statistics are not always reassuring to a sick person. Such data is objective, but the interpretation can be controversial - even among experts. Although patients deserve access to this information, I have found that many patients' eyes glaze over when statistics are quoted. If your patient has a strong mathematical background, consider providing specific references or reprints. I will make some additional observations about communicating statistical data in the special case of clinical trials in chapter 5.

In doubtful situations patients often say, "What would you do for yourself or a family member?" It is understandable that a patient would do so. The doctor is supposed to be in possession of all the facts, perhaps even inside information. But the doctor is a different person from the patient. Beyond this, what one says he will do in a hypothetical situation does not always correspond to what he might actually do when faced with a real situation. Whether doctor or layman, a sick person's priorities, fears, attitude toward risk and hopes for the future are all altered by the experience of illness. For these reasons, unless you are dealing with someone who closely mirrors your own life situation, I do not encourage framing an answer in terms of "what I would do if it were me."

Least helpful is the all too common practice of diligently laying out as many facts, pro and con, as you can muster and closing with words to the effect, "I have discharged my responsibility. It is now your choice to make." This a common practice when the choice is a close call. A doctor might feel that he or she is only being fair or impartial by leaving it up to the patient. But, even when it is a close call, I believe it is part of the doctor's responsibility to lead the decision one way or the other.

If you have taken time to learn something about your patient's values and priorities you may see that these factors tip the balance in favor of one alternative. If you have diligently laid out the facts you have no need to fear that you are coercing the patient. In fact, having a concrete recommendation against which to weigh the alternatives may facilitate the process of decision making for the patient.

One of the great fears common to most sick people is that they will run out of treatment options. The more serious the illness, the more likely a recommendation for treatment will be quickly followed by the question, "What if it doesn't work?" Your task here is to be supportive without being drawn into speculation about future alternatives. First, state firmly that there is always "something to do." But then point out that the more important question is whether any active intervention should be undertaken.

For example, when an illness responds for a while and then becomes resistant it presents a situation very different from when it fails to respond initially. If a patient is not having severe symptoms and has a slowly progressive disease after failure of intensive treatment, a strategy that emphasizes symptom control may be just as acceptable as immediately turning to burdensome second-line therapy. When the patient's response to the initial treatment was a lengthy one, it is also supportive to point out that future advances in therapy may lead to alternatives that do not exist today.

When a patient's concern focuses on the distant future explain that, because it is impossible to know for sure what the situation will be in the future, you cannot make wise choices about the next treatment before the present treatment has even begun. Finally, the choice of subsequent treatment will have to be made by balancing the chance that it will be helpful against the chance that it will cause harmful side effects. These probabilities are also impossible to predict at the outset of treatment. And this leads to a discussion of how to talk about risk.

What Are the Risks?

No drug is entirely free of side effects. Even diagnostic procedures can cause harm. Certainly no surgery is without risk. Whether a patient raises the question or not, in the course of clinical practice a doctor must decide many times a day whether to talk about risk and what to say. A discussion of the specific criteria that must be met to require a discussion of risk in all cases is beyond the scope of this book. There are, however, important principles that should influence what is said when the subject comes up.

Drugs pose a special problem in respect to risk. As recent experience has shown, the fact that a new drug receives FDA approval for a certain indication is no guarantee that all the risks associated with that drug have been identified, let alone characterized as to frequency. My view is that the shorter the period a drug has been in use, the stronger the case for stating that some of the risks of taking it may still be unknown. Conversely, many useful drugs that have been prescribed for decades have serious side effects that occur so rarely as to be of little practical significance compared to their benefits.

Describing these side effects in vivid detail will ensure that some patients will be too frightened to take the medicine. Another category of drug risk involves adverse effects that become a factor for an individual patient only after prolonged use. If a strictly limited exposure is expected for a patient I do not think there is an obligation to describe a host of long-term side effects. On the other hand, a rare side effect that causes serious, permanent damage warrants explicit mention in all cases.

For every known risk of an adverse event, past experience can be analyzed to give an estimate of the probability that it will occur in the future. One can use the language of probability to express risk in medical practice (one in a thousand, 10 percent, a third, etc.). But the prevalence of innumeracy in the general population is high enough to make the use of percentages, ratios and expressions of statistical probability incomprehensible or, at least, confusing to many people (Crowson et al. 2007).

Equally important is the fact that estimates based on statistics are accurate only in relation to populations. Any given individual patient may have higher or lower than average risk because of his or her unique characteristics (Edwards and Prior 1997). Avoiding statistics by making a common qualitative statement (rare, a few, hardly ever, etc.) suffers from this limitation and is likely to be vague or subject to manipulation as well.

Comparing risk to a familiar reference (safer than driving to work, less than the chance of being hit by lightning) has the advantage of being concrete and readily understood. Unless you are willing to collect reliable data about the risk of these common experiences, however, it is not entirely responsible to use them as a reference. The best practice is to be thoughtful about which expression you choose for an individual patient and be prepared to defend the accuracy of your statement if it is ever questioned.

Be aware that how you frame your description of a risk may have a powerful effect on how a patient perceives it. For instance, saying one person in a hundred will experience numbness after an incision is not at all the same as saying 99 percent of people will not have numbness. Mentioning intestinal damage is a far cry from saying that a treatment could cause a hole in the bowel leading to peritonitis. Despite the chance of frightening a patient, I think it is unwise to yield to the temptation to soft-pedal serious complications or side effects. Avoid euphemisms. If you are talking about seizures be sure your patient knows what a seizure is. "Convulsion" might be better understood. If there is a chance of fatal infection do not limit your discussion to "lowered blood counts." "You might die of infection" is a more forthright expression of the reality.

Given the extensive, often melodramatic, but usually superficial coverage of medical subjects in the popular media you will spend a lot of time reassuring patients that the latest medical news does not refer to them. Here, as in the case of describing probability, it necessary to know your facts. If your patient is asking about a report that is unfamiliar to you or sounds garbled defer comment until you do have the details. It is quite embarrassing to have to replace confident reassurance with a recommendation to stop a medicine or forgo a procedure.

The prevalence of direct to consumer (DTC) advertising presents another problem. These ads always cite a list of side effects at the end of a usually glowing portrayal of the drug's benefits. The side effects (sometimes fearful sounding) are presented in rapid fire on radio or TV in the verbal version of fine print. The problem arises when your patient has just received a new prescription for the drug in question and hears about the rare side effects for the first time from the ad rather than from you. It pays to be aware of which drugs are currently being heavily promoted in DTC advertising and take extra care to discuss side effects when you prescribe one.

The Existential Question

I want to express some strong personal ideas about the specific case of a prognosis that includes a quantitative estimate of survival. In twenty-eight years of practicing oncology I met many patients who had, earlier in their care, received a quantitative survival estimate from another physician. I know of no single instance when a patient benefited from such an estimate. The great majority of these people experienced strong, long-lasting negative feelings about the survival estimate, often mixed with anger, depression and panic. What is surprising is that, in most cases, the estimate was given in well-meaning response to a sincere request by the patient. Occasionally the physician, perhaps responding to the modern emphasis on openness and patient empowerment, chose to offer a survival estimate gratuitously. These unsolicited survival estimates invariably led to anger and a loss of trust. In either case the estimate was never helpful.

Seriously ill people have a powerful desire to see the future. Many will ask, "How long will I live?" I believe the correct response to a request for a survival estimate, however urgent or sincere, is "I don't know how long you will live." This is, in fact, the only honest answer. An individual is not a statistic. The statistic of median survival is often used to express life expectancy for serious diseases. We should never lose sight of the fact that, even with the most authoritative data, one-half of patients will not survive to reach median survival and the other half will survive longer, in some cases much longer. Rarely, if ever, is it possible to predict where on a graph of survival versus time an individual patient will fall.

Sometimes a patient will have significant business obligations or other financial issues that will be affected by his or her death. When this was cited as a reason for needing a survival estimate I learned to respond, "My family knows what to do if I do not come home tonight. And I don't have cancer. I think everyone should have similar arrangements."

In many cases the question will be loaded with poignant meaning. Young parents will be thinking about their children. Elderly people will be thinking about their disabled spouse. When faced with obvious heartbreak remember that your role is to provide comfort, not certainty. You cannot promise the young parent long life, but you can say that experts tell us parents have already given most of their good influence to a child in the first few years of life. You cannot change the reality that a dying spouse will not be present to care for the survivor, but you can give the patient information about practical options for community agency or social service support. Resist the impulse to give survival estimates when the real need is for care.

Despite the inevitable requests to do so, it is equally unwise to discuss survival with family members outside the presence of a patient. Family members experience severe distress owing to the uncertainty that surrounds terminal illness in a loved one, almost more distress than they feel because of anticipation of the loss itself. Their requests for survival estimates are an expression of their intense angst resulting from uncertainty as well as their struggle to prepare for anticipated grief and loss. It is an illusion to think that specifying when you expect a patient to die will provide lasting relief from this angst.

Any lessening of uncertainty will be replaced by an increase in dread.

When death seems imminent out-of-town family members will invariably ask, "Should I come now or later?" Since you cannot predict the time of death precisely, tell them, "If you must be absolutely certain to be present at the time of death you must come now. But consider that you may be faced with a longer wait than you can manage and that you might be more help to the rest of the family after death has occurred." In this case, as in all cases, a specific estimate of survival is poor therapy for the distress and, coming from a professional, has the weight of a death sentence. You will nearly always be wrong, either too optimistic or too pessimistic. Like a broken record, respond to both patient and family members that your task is to do what you can to provide the best quality of survival for whatever time the patient has left. This statement has the virtue of being both true and, if not momentarily satisfying, at least harmless.

Answering, "I Don't Know"

I should not close this chapter without commenting further on choosing to say, "I don't know." This choice seems counterintuitive. The pressure to have an answer for everything in medicine is considerable. The media, both through sensational reporting and the increasing use of DTC advertising, contribute to an exaggerated expectation on the part of the general public of what medicine can offer today. Doctors who work in specialties where advertising has become commonplace and medical researchers who promote their work while seeking funding and recognition also contribute to the problem of unrealistic expectations.

My point in mentioning these influences is to remind you that there is still a role for humility in talking to patients. There remains a great deal that we do not know. There will be occasions when it is best to simply say that you do not have an answer. Ideally, this can be accompanied by a promise to find out or a referral to another professional who might have an answer. Sometimes, instead of losing confidence, a sick person can find this refreshing. In all cases, when "I don't know" is the only true response, it is to be preferred over a guess.

CHAPTER FOUR   BAD NEWS

From trivial to disastrous, bad news assumes an endless variety of forms: "You are slightly overweight," "Your cholesterol is too high," "You have high blood pressure," "This rash is shingles," "The chest film shows a spot on your lung," "The biopsy shows a malignancy," "Your treatment has failed." The importance attached to communicating bad news in medicine is illustrated by a search of the National Library of Medicine online database, PubMed, which yields seventy-five papers published in the past four years on the subject of teaching how to do it. Dr. Robert Buckman has written an entire textbook on the subject (Buckman 1992). Because it has been covered so extensively, I address this topic with some trepidation. But discussions of how to communicate bad news in medicine usually center on potentially fatal illness, and a few mundane but more common situations also deserve attention.

How to Approach the Presentation

It is tempting to convey unwelcome information of minor significance without paying much attention to language. But what seems minor to the doctor does not necessarily seem minor to the patient. For someone whose close relative died of a stroke at forty there is no such thing as "mild" high blood pressure. If someone close to you had an amputation because of diabetic complications a borderline blood sugar may be alarming. In all cases the language with which a doctor gives bad news to a patient and the way in which it is expressed can be constructive or harmful to subsequent care.

Don't begin to disclose bad news by putting yourself at a disadvantage. Be sure there is sufficient time for explanation. Choose a setting that is private and free of interruptions where both you and the patient can be seated. Have all the available facts at hand before you start. Few things impair a patient's confidence more than seeing you thumb through your notes, call for more records or draw a blank when a question that you are unprepared to answer is asked. Euphemisms and jargon both increase the risk of misunderstanding. Careless choice of words can cause needless alarm. In other words, think before you speak.

Suppose the result of a blood test shows a slightly elevated creatinine value. While it might be technically true to say this is an indication of "failing" kidney function, words like "failure" are freighted with ominous meaning that would be inappropriate on the basis of limited information. It would be equally unhelpful to simply recite the lab result, which may have no concrete meaning to a layperson. Explaining the significance of a result like this by saying, "It sometimes indicates that your kidneys are working less efficiently than normal," will strike a reasonable balance between accuracy and caution.

When a patient reacts to whatever you have said with undue alarm, resist the impulse to instinctively give blanket reassurance with words like "Don't worry, it's probably nothing." Premature consolation with such phrases sounds trite or false to a hard-nosed realist and will not, in any case, reassure the highly anxious patient. Even if you think it is true, avoid using any language that implies, "It could be worse". Most people find this message unwelcome after getting bad news. Whatever is happening to a person is what is worst for that person at that moment. There is a natural tendency, when communicating information of great import, to push on with your explanation non-stop in an effort to get the whole story out as quickly as possible. But a patient hearing bad news is the one who is most in need of time to comprehend and digest what is being said. The emotional impact of bad news almost guarantees that accurate recall and full understanding will be impaired. This phenomenon will be discussed more fully later.

There is a simple but effective measure for lessening the "my mind went blank" reaction to bad news. In a word, pause.

After hearing distressing information many people will sit in stunned silence, unable to formulate a meaningful question. If you inject several quiet pauses while you are explaining the new information it will allow opportunities for a patient to collect his or her thoughts, ask for repetition or ask a question that immediately comes to mind. It is unfair to rely on the pro-forma closing phrase, "Do you have any questions?" at the end of your presentation. By this time questions that occurred to the patient while you were speaking have fled from his or her thoughts.

An invitation to interrupt with questions before you launch into an explanation is a reasonable measure. When the information is complex or involves evidence from several sources consider jotting a brief outline of the data as you speak or at the end of the meeting giving the patient a concrete record of what was said. Finally, as another measure to combat fear and helplessness, be prepared to outline at least a preliminary plan of action to deal with the problem immediately following your explanation.

***

In devising a strategy for giving bad news it is useful to take a systematic look at two common sources that give rise to it; physical examinations and facilities outside your own office.

Physical Findings

You may find marked elevation of blood pressure in an otherwise healthy person in the course of a routine physical. You may find a mass in the breast, an enlarged liver or a nodule in the prostate when an established patient visits you for another problem. An observation of this kind will be just as unexpected for you as it is for the patient. There will be little time to carefully prepare a way to explain the finding, let alone formulate a plan for further study and treatment. Without an opportunity to reflect at leisure or check your library, you must rely on your fund of knowledge and communication skills to formulate an explanation. Whether we admit it or not, an unpleasant surprise can be as unsettling for the doctor as it is for the patient. Aside from being caught off guard, on this occasion you will not be able to refer to outside sources of information before you have to say something potentially troubling.

Especially in the case of a threatening finding, the physician's own emotional response, over which we have less than perfect control, can alarm the patient. A normally chatty doctor becomes silent. His or her customarily sunny disposition becomes serious. A frown distorts the doctor's expression. Patients tend to be acutely aware of these non-verbal signals. I am not advocating that you attempt to mask these reactions with false bravado. In fact, such efforts will usually be transparent. Simply be aware that your patient will notice that something is up well before you have said anything, will be intently watching your subsequent behavior and may even interject, "What's wrong?"

At this point, the key is not to blurt out the first words that come to mind. If you need to spend more time on the examination, acknowledge that that is what you are doing and ask for a moment to confirm your finding and refine your observations. Avoid a premature announcement of what you think you have found, and do not rush to make immediate efforts at reassurance. Once you are past your initial surprise you may work on projecting thoughtful professionalism and collect as much additional information as possible before explaining the findings.

For findings of greater significance, although it may require deflecting the patient's questions, I think it is to the doctor's advantage to ask the patient to dress and take a seat before launching into a discussion of the finding. If possible, leave the room, do not allow interruptions for phone calls or distracting office problems and take a few moments to collect your thoughts. Now you have armed yourself as well as possible and are in a position to use the strategies discussed below.

Be sure to use lay language to describe physical findings such as heart murmurs, neurologic signs and arterial bruits. Use common recognizable references to estimate the size of lumps and nodules. Words such as centimeters, lesion, mass, ascites, and edema are not helpful here. Drawing a diagram is a fine way to supplement your description when appropriate. Concrete facts tend to reduce anxiety. Ambiguous descriptions tend to increase it. The significance of the finding, in essence the list of conditions that it might represent, is another matter. Your decision about how far to go on this track must depend on your own assessment of how well prepared you are to do this.

Initially your description should usually be limited to the physical characteristics of what you have found. If your description strays into a listing of possible diagnoses at this stage you have probably gone too far. Especially if the list includes serious diseases (to a patient confronted with an unexpected finding this means just about everything), you may open the door to a flood of questions you are not so well prepared to answer. Having to dodge these questions or give vague, confusing or incorrect answers is worse than leaving a patient in suspense about what the finding might mean. There is nothing wrong with saying that you do not yet know what significance, if any, the finding has. Then be sure to add that you expect further studies will clear things up and that you will be completely candid about the results of that investigation.

Outside Results

For the results of procedures done at another site, barring some breach of protocol, the doctor will learn of the result before the patient. In this situation you have a chance to be armed with all the information you need to formulate an effective description of the findings before you speak to your patient. The Internet, your own library and colleagues are all available as resources. Surprisingly, physicians sometimes fail to take advantage of such resources before giving results to a patient. If your office routine allows an assistant to give unscreened reports to patients on the phone, your next call from the patient who has learned of an abnormal result this way may catch you at an awkward time or before you have done the review needed to give good answers to questions.

A wise policy is to have all abnormal results screened by the physician before anyone else provides them to a patient. If the result involves a scan, x-ray, or endoscopy procedure, taking the time to discuss the written report with the radiologist or endoscopist before you try to explain it to a patient will pay big dividends in the quality of your explanation. As with physical findings, diagrams can be invaluable. Having the actual x-ray or scan image available for review in your office when you see a patient will enhance your explanation more than any other measure.

A diagnosis of malignancy or any other condition of similar importance should never be given over the phone unless there is no other possible option. Having a member of your office staff call a patient to "come in for results" without disclosing what the results are carries obviously ominous significance. It follows that when you anticipate a pending test result will contain information of this magnitude, it makes sense to schedule an appointment for results at the time the test is ordered or performed. A patient also deserves to have your recommendation for treatment or referral to a consultant at the same time you tell the patient of the diagnosis.

***

When a patient's survival is in doubt several important aspects of managing bad news go beyond the physician's choice of language.

In One Ear and Out the Other

Truly bad news raises the prospect that a patient will not remember much of what you have said after he or she leaves the office. Psychologists may be able to explain why this happens, but, whatever the cause, it is a remarkable phenomenon. It affects people quite independently of their level of education or sophistication and is very common (Silberman et al. 2008).

Although there may not be a foolproof solution to this problem, it will help to make a handwritten outline of the main points of your explanation as you speak and to give this to the patient at the end of the visit. The outline should include the essential findings leading to the diagnosis, diagrams when appropriate and a list of recommendations of what steps should be taken to deal with it. Some authors have recommended going a step further by providing the patient with an audiotape recording of their consultation with the doctor (Knox et al. 2002; Liddell et al. 2004).

Especially in the setting of a new diagnosis, it is not safe to assume that a sick person will remember the details of your explanation, let alone accurately repeat them to a third party or act promptly and appropriately on your recommendations after an initial conference. As you assess your patient's reaction to the information, be alert for signs of failure to comprehend what you have said. Be prepared to suggest sources of emotional support if it is needed. By all means expect that it will be necessary to repeat what you have said on one or more occasions before it sinks in. Have a concrete, specific plan for follow-up before concluding the visit. For your own protection from liability, have a fail-safe procedure for ensuring that a patient has followed your recommendation or sought care elsewhere in case he fails to return.

It has recently been suggested that giving bad news over the phone might be justified because it gives patients a chance to collect their thoughts and formulate questions before seeing the doctor in person (Ulene 2009). While this might apply to an occasional person with ample emotional resources, I believe that, for the great majority of people, this practice creates the risk of impulsive behavior or exaggerated emotional responses without ready access to support. On a practical level the doctor who gives any but the most trivial bad news over the phone had better be prepared to see the patient immediately. To not do so risks provoking bitter disappointment and anger.

Last, but not least, recognize the impact on your own emotional state of having to give terrible news to a patient you have been caring for. I assume you have made sure that an ample supply of tissue is at hand before you sit down together. If the patient is not the only one who needs to use it, do not fret. A show of honest emotion on the part of a sympathetic physician will be seen as evidence of caring, not a sign of weakness. When you find yourself powerfully moved by what has happened to your patient seek out a trusted colleague with whom you can share your feelings rather than hide them. Suppressing your own genuine sadness can have a corrosive effect on your ability to cope over the long haul.

Whatever You Do, Don't Tell Him

In several cultures there is a tradition of hiding a potentially fatal diagnosis from a sick person. Usually this tradition is based on the belief that learning of the diagnosis will cause an individual to "give up hope" and the corresponding fear that losing hope will hasten his or her death. When caring for a patient with this background you may expect to be approached by a family member who urges (sometimes orders) you not to inform the patient of the diagnosis. This runs counter to the current emphasis in medicine on empowering patients and engaging them in making decisions about their own care.

Hiding a diagnosis is also highly unrealistic, given the obvious, widely recognized significance of certain procedures, diagnostic tests and hospital units. I have found that when an attempt to carry out this deception had been made before I met a patient, he or she usually knew quite well what was happening but willingly went along with the deception out of concern for the family's feelings.

I have always felt that deception of this kind is unfortunate. When it is practiced, aside from the fact that the doctor-patient relationship is based on a lie, the patient and family are both deprived of a chance for meaningful mutual support. Patients in this position are, in effect, forbidden from saying important things they might otherwise wish to say. Furthermore, patients can hardly be expected to participate in treatment decisions without the ability to talk about their own diagnoses. The family, having to maintain the appearance of good cheer, is likewise prohibited from giving the kind of comfort a sick person most values. If you decide to attempt to reverse the practice of deception, the first task is to gain the family's confidence.

For this reason it is a mistake to force the issue of disclosure early in your relationship. When the time is right, after you have established good lines of communication, you can approach the patient's primary caregiver. Begin by explaining what both family and patient stand to lose by continuing to hide the truth. To counter the fear that the patient will give up and die, explain that, in all likelihood, he or she already knows the truth and is simply pretending not to know in order to obey the family's wishes. I have always been skeptical of the commonly held belief that "a good attitude" is critical for a good outcome or, conversely, that pessimism guarantees a bad outcome. I recognize that making a convincing case to the contrary with a worried family is a tall order in this setting, but it is in your patient's interest to try.

The reverse situation can also occur. A sick person, feeling an obligation to protect loved ones, insists that they not be informed of a dire prognosis. As in the previous example, if this instruction is followed, both parties wind up facing one of life's ultimate challenges in a state of enforced isolation. Whether speaking to a patient or a family member, try a gentle reminder that, even if such a deception could succeed, the value of having patient and family united in facing severe illness would be greater than any harm that might come from knowing the true nature of the diagnosis. A simple analogy can sometimes be persuasive; for example, "You would not go away on a long, difficult journey without saying good-bye, would you?" or "What if your positions were reversed? How would you feel if you were unable to talk about the illness with your loved one?"

An aspect of the practice of hiding the truth that is seldom discussed concerns the sick parent who has very young children. Especially in the case of serious illness you may be asked for guidance about how the parent should explain his or her illness to a child. These situations are invariably both sensitive and poignant, and it is an illusion to think that you can provide the parent with a specific script. There are, however, some observations that may help a parent who finds it difficult to say anything.

First, the parent should be reminded that children are keenly aware when their parents are in distress, even when nothing has been said. Second, young children may find the actual facts less alarming than the version conjured up by their vivid imaginations. Finally, children need to hear that illnesses just happen by chance and that nothing they did or failed to do contributed to it. Choosing the right words for a child whose parent is not expected to survive will tax the ability of even the most experienced clinician and may be a task best reserved for clergy or a child psychologist. I do not pretend to have easy advice to give on the subject.

Who Will Carry the Ball

For some medical problems you will choose to be the primary physician. When the problem is simple you may prescribe treatment without further ado. For more complicated findings you may have to order the tests or procedures required to reach a diagnosis. For a potential problem clearly outside your own expertise you may have to make a referral to a specialist even before ordering additional tests.

When a specialty referral is appropriate you must either rely on your own fund of knowledge to decide which tests are needed and order them yourself or arrange a consultation as the first step in investigation. I suggest that when the unexpected finding represents a potentially complex problem, unless the tests that are required are obvious, you should consider explaining to your patient that the consultant is in the best position to guide the diagnostic workup in the most efficient way.

Many patients will accept a referral to a specialist without question. Others will need to be assured that you have chosen "the best possible doctor," or words to that effect. Your response will, of necessity, depend to a great extent on the setting in which you practice. In an urban setting you may be able to choose from a number of highly qualified local physicians. In rural areas your choices may be limited to out-of-town centers. For patients with some forms of insurance, you may have to select from a limited panel of specialists.

In any case, you will usually have to admit that, however well qualified your first choice for a referral is, it will be difficult to certify that specialist as "the very best." It is better to say that your confidence in the specialist is based on your own firsthand experience than to pretend that it is possible to determine who is "best." It is useful to know some details of your specialist's curriculum vita and be prepared to mention these.

In spite of your advice, a patient who is motivated by a powerful need to be convinced the specialist is somehow exceptional will often take it upon themselves to identify an expert affiliated with an academic center, recommended by a friend or relative or named by some authoritative source. Endorsements of this kind have great appeal to anxious sick people. And they will outweigh other factors. Your confidence in your own recommendation will not have much influence in this situation. Some special considerations apply when patients inquire about referral to an out-of-town academic center.

I found it useful to describe my own criteria for referrals to distant centers to patients who are considering such travel and ask for my advice. A medical center's reputation for excellence is not enough to justify long-distance travel for some patients. The costs, both economic and emotional, of dealing with serious illness away from home are substantial. Travel involves loss of continuity, time and the disadvantage of being far from family support and entails considerable expense. And a reputation for excellence in medical research does not always equate to excellence in the way a patient's needs for personal care and communication are met.

Therefore, at least one of several conditions should be met to fully justify traveling to a distant medical center. First, the patient's condition may be so uncommon that only the medical center has sufficient firsthand experience in management to be able to do it well. Second, there may actually be a genuine superstar physician who manages complex procedures better than anyone else. Third, progress in therapy for the specific condition may have advanced so rapidly that the latest developments have not yet made their way into community practice. In these cases, referral to a center may be lifesaving. In ordinary circumstances it may only be an unnecessary burden. When this explanation fails to reassure a patient about the wisdom of receiving care in his or her own community, as it often will, it is best to quickly acquiesce and be as helpful as possible in making a prompt and well-documented referral to the center that best fits the patient's diagnosis and ability to travel.

Consultant versus Primary Physician

When you are speaking as a consultant rather than a primary physician some special principles apply. Most obvious is the fact that, as a consultant, you have obligations to the referring physician as well as to the patient. The first of these obligations is respect. Shocking as it may seem, I have known of overworked consultants who react to a referral by complaining to a patient, "Why were you sent here? Your problem belongs in some other specialty." Or worse, "Your doctor could have taken care of this himself."

In one fell swoop the referring doctor will have been insulted and a patient needlessly dismayed. The same result is achieved when a remark is made to the effect that a referring doctor missed an easy diagnosis or prescribed an incorrect treatment. Except in cases of clear-cut incompetence or professional misconduct, a consultant will best help a patient by supporting his or her confidence in the referring physician and only harm a patient by degrading that confidence.

A subtle problem for consultants arises in relation to communicating their opinions to the referring doctor. In an earlier era, when medicine proceeded at a more formal and leisurely pace, it was considered very bad form for a consultant ever to give specific advice about diagnosis or treatment directly to a patient. Rather, the consultant was expected to make his or her report to the referring physician, who would then communicate the consultant's findings to the patient and carry out the recommended plan. Obviously, this style of practice is now, for practical purposes, obsolete.

In a era when fax machines, e-mail and cell phones are taken for granted, patients are not likely to be satisfied with waiting to learn what a consultant thinks until they hear from their primary doctors. But it remains awkward and embarrassing for a primary physician to receive questions from a patient about a consultation before seeing the consultant's report. There is a way to preserve the mutual respect between consultant and referring doctor that once graced the practice of medicine and still meet the demands of modern patients.

This can be achieved if referring doctors explicitly give the consultant permission to discuss findings with a patient when making the referral. Consultants, in turn, should scrupulously follow the practice of calling the referring doctor with at least a preliminary report immediately after seeing a patient. If all physicians observed these rules a great deal of avoidable friction between primary doctors and specialists would be eliminated and sick people would benefit.

Honesty versus Hope

Most advice about how to communicate life-threatening new information ultimately evolves into a discussion of how to describe prognosis. Although conveying reassurance was discussed earlier in relation to answering questions, I would like to emphasize the importance of hope in the setting of incurable illness. While I continue to be skeptical about any evidence for a biologic influence attributable to a patient's attitude, I have no doubt that hope is a precious and fragile asset for any sick person. One's frame of mind affects physical performance, appetite and personal relationships. No one should ever intentionally or inadvertently speak to a patient in a way that destroys hope. Sadly, in contemporary America, the well-intended emphasis on patient empowerment, demands for full disclosure and criticism of paternalism in medicine lead some doctors to do just that.

During my years in the practice of oncology, a number of patients came to see me after leaving the care of a competent and caring physician to seek care elsewhere. Many of these patients made a decision to change doctors because they heard statements that left them, at least momentarily, without hope. And virtually every patient I saw who had abandoned standard care in favor of unconventional therapy or quackery did so for the same reason. I will have more to say about dealing with medical quackery in the next chapter. Feelings of anger and bitterness result when a sick person interprets a doctor's words as an expression of hopelessness. This happens without any relation to the objective accuracy of what was said.

Unfortunately, it happens even when the doctor made such a statement with the best intentions. Why should this be? I think it is because a fundamental characteristic of the human condition is the need to have something to look forward to. When our very survival is threatened by illness, the need to believe that things just might get better becomes very powerful. When hope is taken away an individual's need to restore it can overcome rational thinking, logic and even common sense.

Those who dogmatically promote physician candor and patient autonomy in medical practice often have not had firsthand experience treating people with severe illness. In a sincere effort to respond to calls for candor and patient autonomy, doctors sometimes feel compelled to be brutally honest in describing a diagnosis as incurable or fatal, to quote grim survival statistics or to use language that explicitly or implicitly says there is no treatment for the patient's condition. If one accepts the validity of both the vital importance of hope and the objectives of candor and patient autonomy, how does one deal with the resulting conflict?

First, as I have already urged, do not make survival estimates. Nothing else limits hope as certainly as predicting when you think someone will die. Second, when starting treatment, always remember to mention that alternate choices and backup measures are available in case of a treatment failure. Third, especially for diagnoses that are typically slowly progressive, emphasize the promise of biomedical research for developing new treatments in the future. Fourth, when survival is likely to be short, use your knowledge of an individual patient to identify an achievable goal that has meaning for that person - perhaps an upcoming holiday or a much-anticipated trip, an anniversary, graduation, expected birth or some other family milestone. Encourage patients to believe that you will help them try to reach that goal and that if they reach it you can work together to reach another goal beyond the first one.

A deft touch is required to do this well. Sick people have different priorities, and some bargaining may be needed to settle on a realistic goal. But a corollary benefit is that they may wind up with a more accurate perspective on their prognosis without feeling that hope has been dashed. When severe physical problems limit the range of realistic goals, it may be that all we can offer is relief of a specific, troublesome symptom or temporary improvement that will permit return to a favorite activity. A goal, however modest, is still something to strive for.

A more complex situation to consider is the one that results when, after failure of multiple treatments, remaining alternatives are more likely to cause unwanted hardship and toxicity than meaningful benefit. In this case the physician knows that a decision to continue aggressive therapy will harm more people than it helps. Explaining this situation with honesty and sensitivity is among the most demanding tasks a doctor can face. But the advantage in this situation is that you will be talking to someone who has a good deal of firsthand experience with both the illness and the costs of previous treatment. Such patients will be under no illusions about the burden of side effects. The disadvantage is that they are, if anything, facing an even more immediate threat to their survival than they did at the time of the initial diagnosis.

Your obligation is to suggest, using all the skill and care you have to offer, that quality of life is sometimes to be valued more than length of life and to promise to do everything in your power to provide quality even if the patient elects to decline specific therapy. Toward the end of life many sick people experience feelings of resignation and acceptance. This change in emotional state and simple exhaustion make a decision to suspend treatment other than relief of symptoms easier for everyone to accept.

However, as serious illness advances, most people who are still relatively active have instincts for survival that are operating at full strength. Or a patient who might otherwise decline further treatment feels an obligation to keep trying in order to satisfy their family's expectations and hopes. Be aware that, once a treatment option is identified, the great majority of sick people will find it extremely difficult not to try it.

Whatever influences are in play, even when the chance of benefit is vanishingly small, only a few strong-willed individuals will resolve the doctor's dilemma by making a firm decision to decline further aggressive therapy. When your judgment leads you to believe the patient should choose supportive care your own ethical principles must guide a decision whether to refer the patient to someone who will offer the last-ditch option or, with some reluctance, to carry it out yourself. If you do pursue treatment you must be capable of putting aside your reservations and devote your best energy to managing care even when you wish the treatment had been declined.

CHAPTER FIVE

EXPERIMENTAL AND UNCONVENTIONAL THERAPY

Clinical Trials

Enrollment in a clinical trial is an option for patients with many serious diseases when there is no "standard" treatment or when standard treatments have been unsuccessful.. Clinical trials involving human subjects in the United States must be approved by the Food and Drug Administration. In these trials, patients are selected according to predetermined criteria and offered treatment with a promising new drug. After careful, sometimes lengthy, follow-up, the results of treating these patients are collected and analyzed in order to evaluate the safety and effectiveness of the experimental therapy.

Clinical trials are classified according to the stage of development of the experimental treatment being studied. In a Phase One study a relatively small number of patients (usually fewer than one hundred) are treated with a new therapeutic agent that is being administered to humans for the first time. A Phase One clinical trial may aim only to study how the drug is metabolized and its tissue distribution. Or it may be limited to defining the optimal dose of a new agent - the minimum dose with any therapeutic effect and the maximum dose that can be given without unacceptable toxicity. In cancer research, when results of these preliminary studies are available, another round of Phase One trials will be conducted aimed at measuring the effectiveness of the new agent against a variety of malignancies in the hope of finding useful activity that is specific for one or more of these specific diagnoses.

Once an experimental therapy has demonstrated sufficient effectiveness and safety in early trials to justify further study, Phase Two studies are designed. In these studies it is often necessary to enroll patients at several locations in order to accumulate large enough numbers to provide convincing statistical evidence of effectiveness and safety. In contrast to Phase One trials, in which the new treatment was used alone, it may be given in combination with one or more standard drugs.

When the trial compares treatment with the new agent to results achieved with one or more standard treatments it is referred to as having one or more "arms." The control arms employ a treatment regimen chosen to represent the best available standard treatment. A significant feature is that patients who participate in a multiple arm study will be assigned to one or another of the treatment arms by a process called *randomization*. In this process, analogous to drawing cards from a deck or flipping a coin, a computer, programmed to make random assignments, designates which arm of the study a particular individual patient will enter.

In some cases it is appropriate to choose to compare an experimental treatment to no treatment at all. In such studies patients assigned to the control arm of the study are observed without any active therapy. Multiple arm studies raise concern that knowledge of which patients receive the experimental treatment rather than a control arm will influence interpretation of the results. Optimism about the effectiveness of new treatments can have a powerful effect on how sick people feel in the absence of any objective benefit. Enthusiasm about the prospects for a novel therapy can even influence how professionals interpret x-rays or other supposedly objective data.

The potential for this kind of bias is very relevant in cases where relief of a symptom (e.g., joint pain) rather than an objective measure (e.g., length of survival) is the end point. In order to eliminate this kind of unconscious bias the study can be conducted so that neither the treating doctors nor the patients know which treatment arm individual participants are receiving. Such a study is referred to as a "double blind" study.

Double blind studies require the use of an inactive substance or placebo to replace the experimental drug for patients in the control arm of the study. Through the use of a coding system only the study sponsors know which treatment each patient is receiving. To ensure that an inferior treatment arm is abandoned as soon as it is recognized, ethical principles require that the study design must include a system for analyzing results periodically as the study progresses.

Rules are established in advance to "unblind" the trial as soon as one treatment arm emerges as significantly superior to any other so that no additional patients are assigned to an inferior treatment. In diseases with a typically long clinical course treatment assignments may remain coded for months or years of follow-up. In such cases, even when treatment assignments are not coded, study results are not published until a statistically significant difference between treatment arms is observed.

In a Phase Three study, the most mature form of clinical trial, as many as several thousand patients may be enrolled, nearly always requiring the cooperative efforts of doctors at institutions scattered across the country. In these studies finer details such as different dose schedules or a comparison of three or four different combinations of drugs are analyzed to determine if such manipulations can improve on the favorable results of earlier studies. When one treatment regimen proves superior in safety or effectiveness to other regimens in Phase Three trials, the difference is often small, a matter of a few percentage points of superiority.

Considering the complexity described above it is not surprising that a minimum of several years is required for a drug to progress from experimental to approved, standard treatment. An entire industry has developed to support the hugely expensive administration, data collection, safety monitoring and statistical analysis involved in the process of new drug development. In recent years some pharmaceutical companies have come under criticism for conducting trials called "post-marketing" studies, in which drugs that have already been approved for some indication are supplied free of charge to physicians.

Doctors, usually in community practice, use the drugs in treating their patients and are asked to submit information on their experience with the treatment. Critics of these studies maintain that their real purpose is to encourage familiarity and greater use of the product rather than to gather scientific information. Some observers have criticized relationships between the pharmaceutical industry and physicians in which the physicians are compensated for enrolling patients. The use of these studies can be defended by pointing out that they provide access to useful new drugs free of charge and, more important, that a very large number of patient experiences are required to reveal rare side effects that might been unrecognized in clinical trials conducted prior to FDA approval.

An essential feature common to every legitimate clinical trial is the Patient Consent Form, which must be presented to every patient recruited for the trial (Brown et al. 2004). It must include a forthright presentation, in lay language, of the scientific goals, treatment procedures, and potential benefit, if any, associated with the study. A detailed listing of the potential side effects and risks of all drugs and procedures in the study, along with an estimate of the frequency of these risks, must be included. Alternatives to entering the trial must also be identified.

In FDA-approved clinical trials the consent form is prepared by the organization that sponsors the trial and then reviewed and approved by an independent organization, the Institutional Review Board (IRB). The IRB is responsible for ensuring that the rights and safety of participants are protected. Enrollment can proceed at a study site only after the IRB approves the study itself as well as the Consent Form. As a study progresses and new information about safety is recognized, the consent form must be amended to include the new information.

In theory this procedure should guarantee a measure of benevolent oversight and guidance for sick people who choose to enroll in a clinical trial. Efforts are made to ensure that the consent form is presented in a noncoercive setting and prepared in language that would be clear to any layperson.

The dilemma for a doctor trying to guide a patient through a consent form is that experimental treatments, by definition, involve unknown as well as known risks and benefits that are, by definition, as yet unproven. Further, some studies investigate only pharmacology or physiology and not the efficacy of new treatments. In these trials, although future patients might benefit from the information gained, there may be no prospect of benefit for participants in the study.

It should be apparent that discussing a clinical trial is a far from simple task (Eggly et al. 2008). The doctor's role as counselor is just as crucial here as it is in standard care. If anything, the complex issues associated with clinical trials make effective communication more difficult than when dealing with standard care (Epstein et al. 2004). As with standard care, I believe there is an obligation to give advice about whether the balance between risks and potential benefit favors enrollment in a clinical trial for a specific individual. When suggesting enrollment in a clinical trial, one should be aware that the act of presenting this option implies a bias in favor of enrollment. A doctor should not suggest a clinical trial unless he or she has reviewed the study protocol or, at least, discussed it with the principal investigator at the local study site.

Statistics

Statistical analysis of the results of clinical trials has governed the advance of medical therapy for decades. It can show whether the differences in results observed in a clinical trial are more significant than what might occur by chance alone. Without this discipline we are left with anecdotal reports, personal testimonials and promotional materials to evaluate new treatments. Explaining statistics in relation to clinical trials is both more important and more difficult than it is in ordinary practice.

Even with a firm grasp of the scientific rationale for the design of a clinical trial, it can be difficult to present the statistics supporting the conclusions of that trial fairly to a nonscientist. One reason for this is that expressing the value of a new treatment compared to some standard treatment with a statistic in some cases exaggerates that value. This occurs most often when relatively few patients - for example, one in ten - show a favorable response to the standard treatment

Suppose that a clinical trial showed that two patients out of ten respond to the experimental treatment. It would be statistically true to say that the new treatment doubled the response rate of the standard treatment. But, in fact, in this hypothetical example, only one future patient out of ten would be expected to benefit by choosing the experimental over the standard treatment. This is because the trial results predict that in the future seven of every ten future patients would not be helped by either treatment, two of ten would respond if they choose the experimental treatment and one of ten would do just as well if they choose the standard treatment. In this case it would be true but misleading to simply frame the clinical trial results by saying that the new treatment was "twice as effective" as the standard treatment.

This same statistical quirk applies to preventative treatments in situations in which the risk of a specific disease is low to begin with. For example, in certain situations, women who have just completed primary treatment for breast cancer face a 10 percent long-term chance (risk) that the cancer will recur at some time in the future. With current technology we cannot be certain, immediately after primary treatment, which women are already cured and which are destined to have a recurrence in the future, sometimes far in the future. Thus it may require five or ten years of follow-up to reveal the results of a clinical trial of preventative treatment.

Now imagine a hypothetical new treatment aimed at preventing recurrence of this cancer. And imagine that, after lengthy follow-up, the new treatment is shown to reduce the frequency of recurrence to 5 percent rather than the expected 10 percent.

It would be statistically true to say that the new treatment has reduced the risk of recurrent cancer by half compared to standard treatment. But, looked at another way, one must treat one hundred women with the new preventative therapy in order to help only five. This is because ninety members of the initial group would not have experienced a recurrence of breast cancer even if nothing further were done after primary treatment. They cannot benefit from treatment because they did not need it in the first place. Of the remaining ten, only five will benefit by being spared a recurrence because they received the experimental treatment following their primary treatment. The remaining five did not benefit because they experienced recurrent cancer in spite of the experimental treatment.

If the experimental treatment happens to be associated with a risk of serious side effects, a benefit of this magnitude may be canceled by the risk of the treatment itself. In studies of preventative treatment a result of this kind would not be unusual in clinical oncology research today. It is obvious that a discussion of statistics like these demands considerable skill and patience.

"I Heard About It on TV"

Increasingly today patients are investigating treatment alternatives on their by own using the remarkable resources of the Internet. Every pharmaceutical company, every academic medical center and nearly every community hospital in the nation has a Web site. To a varying degree none of these sites are entirely impartial. Unfortunately, myriad unproven treatments, folk remedies and outright fraudulent treatments also use the Internet to promote their benefits. The Internet is a veritable zoo of good, bad and indifferent information (Mathews et al. 2003). Sites that promote quack remedies may be difficult to distinguish from legitimate sources of authoritative information. Indeed, virtually every example of medical quackery is represented by a Web site, often in a slick and persuasive format. The traditional media, although more reliable than the Internet, often report promising new treatments with great enthusiasm when they are at an early stage of development, far from any practical application. The value of an innovative treatment may be exaggerated or even sensationalized. The details given may be incorrect or so incomplete as to render an opinion impossible.

Whether sick people hear of a promising new treatment in the press, on television or on the Internet, they commonly ask their physician for an opinion. Invariably, the doctor will not have ready access to a reliable, disinterested source for data when a patient asks about a treatment presented in this fashion. Nevertheless, however outlandish the therapy sounds, a knee-jerk negative reaction to such inquiries is to be avoided.

Throwing a wet blanket on the enthusiasm of a worried and hopeful patient will cause unnecessary disappointment and is likely to be resented. Ideally, you can offer to make a phone call and evaluate the source. Or research the treatment yourself in order to obtain enough information to support a comment or recommendation. But, given all the other demands on physicians' time, the question of how far to go to support a patient's search for new treatments is a vexing one.

When the patient can only provide the name of an obscure therapy it is fair to defer a recommendation unless the patient is willing to do the legwork required to identify the source that brought the treatment to his or her attention. At a minimum it is advisable to at least listen to the description and promise to review some objective information and respond at a later date. When that review confirms your initial skepticism make an effort to explain the data as dispassionately as possible without any hint of triumph or ridicule. In fact, it doesn't hurt to applaud the patient's efforts to assist in searching for better approaches to care.

Quackery

Many of the therapies that come to the attention of severely ill people today fall into the category of quack medicine. While professionals tend to scorn such treatments it is unwise to ignore their appeal to worried and frightened laypeople. Living under the threat of disability or death leads to a willing suspension of disbelief, even in ordinarily well-informed, cautious individuals. In an effort to be fair and nonjudgmental it is customary today to describe quack remedies as "unproven" or "unconventional" unless the treatment has been conclusively shown to be fraudulent. In response to public demand the National Institutes of Health established the National Center for Complementary and Alternative Medicine (NCCAM). The NCCAM supports research and provides information on alternative medicine. Its website is the best source for patients seeking scientifically based information on specific therapies.

It is common, after a diagnosis of cancer or other life-threatening illness, for unsolicited pamphlets and promotional material advocating some unconventional therapy to appear on a patient's doorstep or in the mail. Well-meaning friends and family members contribute additional material, usually with the urgent message, "You must look into this." The frequency with which this happens and the fervor with which the material is presented leads me to wonder if part of the motivation for these efforts is the fear that serious illness provokes in healthy people. Loved ones have a need to believe their family member can be helped. Even strangers can identify with a sick person and find comfort for their own fears in the idea that a panacea is out there somewhere if you just look hard enough.

The essential difference between proven or legitimate experimental treatment and unproven or unconventional therapy rests on the significance of evidence. Proven therapy has been shown to have measurable effectiveness and safety in actual experience. The evidence for effectiveness takes the form of accurately recorded data that has been subjected to honest analysis and genuinely independent review.

Legitimate experimental treatment has a rational basis in biologic fact. Clinical trials are overseen by rigorous patient protection procedures. In contrast, unproven therapy is usually supported only by sincere but unverifiable testimonials, anecdotal experience, folktales or pseudoscientific theories. Nevertheless, to these somewhat pedantic arguments, millions of Americans answer, "So what?" as they continue to vote with their feet and seek unconventional treatment. It is no accident that these remedies flourish in those areas where modern medicine, in spite of its advances, still offers treatments of limited value.

Staying in the Ball Game

What is the appropriate response when a physician is confronted with a patient who is considering or has chosen unconventional therapy instead of or in addition to standard therapy? The dictum "Do no harm" is not a bad way to begin. Recognize that it is not useful to attempt to prove the negative proposition that a treatment "does not work." It has been my experience that the psychology that prevails when a sick person chooses unconventional therapy is profoundly resistant to the appeals of logic and even common sense. Above all, the unconventional treatment offers hope. Arguments that interfere with emotional defenses based on hope will usually be perceived as unhelpful, even threatening.

An important element in the promotion of unconventional therapy rests on planting the notion that doctors and "the medical establishment" are uninformed, obstinately resistant to change or involved in a self-interested conspiracy to suppress unconventional therapy. Once adopted, these notions are stubbornly held, nearly impossible to dislodge by a representative of the establishment. A vigorous defense of standard treatment or a spirited attack on unconventional therapy by a physician usually stiffens resistance and plays into the deceptions and delusions that underlie the decision to turn to unconventional treatment in the first place. Therefore, the best course, although frustrating, is to bend over backwards to not be confrontational when talking about this subject. To do otherwise is likely to damage your ability to help.

But the physician has an obligation to point out any obvious harmful consequences of the unconventional treatment that affect a patient (Eisenberg 1997). Simply reciting a list of potential problems, on the other hand, usually falls on deaf ears. The Internet, which often opens the door to unconventional options in the first place, can be a source of reliable information concerning unconventional therapy. A privately operated organization, the National Council on Health Care Fraud, maintains a Web site that contains information about unconventional therapy and has been listed as a reliable source by the American Cancer Society. The Web site, Quackwatch.org, also has an excellent format and offers a wealth of information. Simply entering "medical quackery" in your favorite search engine will yield other good sources of objective information.

If the lure of unconventional therapy has not completely captured a patient's imagination, investigating a particular remedy on one of these sites may help them to make a genuinely informed choice. You might try pointing out the warning signs that identify promotional material for quackery: emphasis on vitamin deficiency, special diets and bizarre nutritional theories, attributing a wide variety of illness to various toxins, treatments that aim to cleanse the body of toxins, success rates that are too good to be true, pseudoscientific jargon, sales pitches for some product, dramatic patient testimonials and claims of persecution by various authorities.

Never underestimate the power of the willing suspension of disbelief generated by the combination of fear and the need for hope that accompanies an illness that is potentially fatal. Too often, in spite of your best efforts, these well-meaning suggestions will be ignored. When this happens and you have a well-established relationship of mutual trust, try to turn your patient into his or her own safety committee.

Do this by asking your patient to evaluate the utility of the unconventional treatment by using the same tests of safety and effectiveness he or she would apply to standard treatment: be alert for unpleasant side effects, and set some measurable goal he or she expects the treatment to accomplish. After a predetermined interval the patient should ask himself or herself if the unconventional treatment has been safe and effective enough to justify continuing with it. In other words, "If you see your condition has not measurably improved or is getting worse, why continue?" In summary, dealing with a patient who turns to quackery demands respect for this person's efforts to help himself or herself and understanding of the psychology of severe illness. And this must be done while exercising forbearance and tolerance in the face of what seems like deception and fraud.

CHAPTER SIX   **INTENSIVE CARE**

The intensive care I wish to discuss is not the kind that depends on ventilators, monitors or artificial hearts. It concerns talking to the sick in ways that lessen distress, strengthen resolve and build trust and confidence. There are several practices that can accomplish this.

Presence

Presence is such an obvious component of being supportive that it tends to be overlooked. Of course you are physically present. But are you engaged and attentive or preoccupied and impatient to move on? Supportive presence requires active listening and concentration. Tight scheduling, faxes, cell phones, beepers, computer screens and Blackberries all compete for attention. Sick people recognize when their doctors are genuinely present and value it highly. Inattention is noticeable and unwelcome to an equal degree. Even devoting only a portion of a patient visit to intense focus on what a patient is feeling and saying will result in increased patient satisfaction.

Distractions are not the only problem facing physicians today. Students are commonly taught or shown by example that it is necessary to maintain a measure of objectivity and professional detachment if one is to take care of sick people effectively. The requirement to be actively involved with a patient's feelings runs counter to this idea. A degree of objectivity has value, but the doctor should attempt to strike a balance between being objective and appearing remote. Striking just the right balance demands some finesse and is a skill that has to be acquired with experience. It is a shame that no one has invented a meter that can warn the doctor when he or she crosses the boundary between being objective and being distant. Being conscious of the importance of balance and monitoring your performance is the crucial first step in achieving this skill.

Validation

In the medical context validation consists of persuading patients that their feelings about being sick are legitimate, the normal result of hardship and travail caused by illness. Sick people sometimes express feelings and ideas that are difficult to listen to. When things are not going well you may encounter anger, pessimism, self-pity or despair. The doctor's task is to make it clear that such emotions are not an indication of weakness or lack of character. The natural instinct to be helpful tends to lead doctors to respond with some form of reassurance that minimizes or even contradicts the patient's feelings. Avoid statements that, in effect, deny the reality of the feelings your patient is describing.

For example, it is better to say "I'm sorry things have been so difficult for you" than to say "Things are not as bad as they seem" or "Joint pain can be a heavy burden" instead of "It's only arthritis." I have also found that any response that implies that a patient is exaggerating does not play well when he or she is offering a sincere expression of some painful condition or loss of function. Reassurance, which works better in reference to what might happen in the future, fails when something has already gone terribly wrong. Conversely, when a doctor acknowledges the existence of an unpleasant reality it lends dignity to a sick person's feelings about that reality.

Many people with severe illness are already laboring under the prevailing emphasis in our society on maintaining a "positive attitude." How often do we hear, "I know he is going to recover because he has such a wonderful attitude" or "He put up a great fight right up to the end," as if to suggest that one keeps going only as long as one fights? But, in fact, reality trumps a positive attitude. Pessimists often get well and optimists often do not. And it is all too easy for a patient to conclude that if a positive attitude helps one to recover, then negative feelings must harm one's chances for recovery.

The last thing sick people need is to feel guilty or fearful when they experience (as they inevitably will) temporary depression, pessimism or worry. Instead, when you recognize this phenomenon, you need to give firm reassurance of the simple fact that the patient's illness does not know what he or she is thinking. When a patient expresses such feelings you need to explain that every sick person has spells of pessimism, fear, and so on, and, more important, that anyone can get well in spite of these feelings.

The emotion of anger is especially resistant to encouraging words. The best framework for responding to anger rests on recognizing that there is a legitimate basis for feeling angry when your life is turned upside down by illness. It can help to state explicitly that you recognize the basis for a patient's anger and that anyone could be excused for feeling bitter about what has happened. The language used to convey this message should be chosen carefully. The goal is to show your respect for the patient's state of mind without adding fuel to the fire. It is not necessary or helpful, for example, to reinforce these feelings by explaining that life is sometimes unfair or cataloguing all the additional reasons to be upset. Just listening, a moment of quiet reflection and a simple expression, "Yes I do understand exactly why you feel that way," is enough.

Empathy versus Sympathy

Sympathy is another natural instinct that arises in the face of suffering. I think the special role of the physician as healer makes empathy a more appropriate response than sympathy. Why make such a distinction? The dictionary definition of sympathy is "feelings of pity and sorrow for someone else's misfortune." Sick people are not looking for pity and sorrow from you. They want you to make them well. They get plenty of pity and sorrow from friends and family members. Patients often told me that what they wanted most from others was to be treated the same way they were treated before they got sick. Empathy, in contrast to sympathy, embodies a cognitive skill, the ability to understand and share the feelings of another, essentially seeing oneself in the other person's place.

The operative word here, at least as it pertains to medicine, is *understanding*. This, rather than the emotional response of sympathy, is the quality sick people want from their doctors. One can feel sorry for another person without knowing much about the details of their problem and then go on about one's business without taking action. Empathy requires a deeper level of insight into what a patient is thinking and feeling and can lead to a more appropriate choice of action to make things better.

In urging doctors to cultivate empathy in clinical practice, I must admit that this is not easy. Empathy conflicts with the need for professional detachment. Doctors have to

114

exercise a measure of detachment if they are to make objective decisions and guard against emotional overload and burnout. One author, in discussing the apparent contradiction between these values, suggests "engaged curiosity" as a better description of clinical empathy and emphasizes the importance of recognizing one's own feelings accurately (Halpern 2007).

For a more pragmatic view of the difference between sympathy and empathy, take the example of caring for a patient with a severe chronic pain syndrome. Perhaps, even though the pain is severe, objective findings are few, diagnostic studies have been unhelpful and the physiologic basis of the pain remains obscure. Approaching such a patient with unalloyed sympathy eventually leads to using analgesics in increasing amounts and potency as the only form of therapy. The potential hazards of long-term analgesic use, including dependency and drug-seeking behavior, are well known. An empathetic approach, while more demanding in terms of time and emotional energy, at least leaves the door open for a partnership with the patient in seeking deeper insight into the origin of the chronic pain and nonpharmacologic means of dealing with it.

Perhaps the most important reason for distinguishing empathy from sympathy is the fact that physicians are often in the position of asking quite a lot from their patients. We routinely ask sick people to exercise forbearance with tedious scheduling, to endure painful procedures, to have patience with the administrative aspects of modern medicine, to accept unpleasant side effects and face the chance of treatment failure. We expect them to think rationally and remain calm in the face of these and other trials. Pity and sorrow seem out of place when placing such demands on a sick person. Understanding does not.

Humor

Humor may seem out of place in a discussion of intensive care in the setting of illness. It does, however, have a uniquely humanizing quality that is of enormous value when it occurs with genuine spontaneity. I do not suggest that you collect a file of good jokes to be told in the clinic or hospital. Similarly, it is not a good idea to try to be funny when you are not a natural comedian. Instead, when something truly amusing happens or is said while you are with a sick person, even when it is in the middle of a serious discussion or a time of sadness, don't be afraid to laugh. Laugh especially hard if the joke is on you. Obviously, don't laugh if a third party makes a joke at the expense of someone else. Laughter is infectious, and you can catch it with no worry that it will make you sick. The critical skill needed in this regard is to be able to see the humor in ordinary events. Laughing at one of the ridiculous things that happen in the course of an otherwise complex medical interaction is okay. It will not impair your image as a professional. Realize that sometimes life is so absurd there is nothing left to do but laugh.

The Optional Contact

Here I want to reemphasize how valuable simple gestures can be for cementing a patient's confidence and trust. One such gesture involves a personal contact that goes beyond the "standard of care." This contact does not have to be elaborate or involve a significant expenditure of time, effort or money.

A phone call at the end of a day to ask, "How are you doing?" after a procedure rather than assuming all is well because you have not heard from the patient is a perfect example. This can be done with a minimal use of time if you establish a routine that calls for your assistant to provide a tickler file with the name and phone number of every patient who has had a procedure on a given day. My own dentists and veterinarians have done this for years. I can't explain why more M.D.'s do not.

If you find yourself in the hospital attending to another matter late in the day, seize the chance to drop in for a second visit to a patient who, although seen earlier on rounds, happens to be nearby. Even if the medical issues were already addressed on rounds this visit is a chance for a brief personal word or an answer to the question your patient forgot to ask earlier. The fact that your presence is spontaneous and voluntary will be noticed and greatly appreciated. If you are already familiar with an associate's patient take the opportunity to at least say hello when you visit your own patient if they happen to share a room.

We fail to realize how positively sick people perceive simple contacts of this kind.

Last, I want to mention a gesture that belongs to another era altogether, the handwritten note mailed in a hand-addressed envelope. The occasion may be as simple as a response to a small gift from a patient, as practical as the report of a routine test result or as profound as sympathy for the loss of a spouse. The very fact that handwritten correspondence is rare in society today endows a gesture of this kind with special significance.

The Impaired Patient

With a severely impaired patient, simple touch is the first and most crucial step in demonstrating that you care. Do not fail to take the hand of your patient at the beginning of a visit. Say hello even when it is not clear you are being heard. Stroke the patient's forehead it if seems appropriate. Make some similar gesture at the end of a visit rather than turn away without recognizing his or her presence.

When caring for patients who are physically or mentally impaired physicians may fall into the practice of turning their attention to and directing their communication to a caregiver or third party. This practice, in effect, denies not only the autonomy but also the humanity of the impaired person. The more severe the impairment, the stronger the temptation to bypass the patient in favor of an able-bodied individual who happens to be in attendance. Resist this temptation. Make an effort to communicate with the impaired patient. This effort is important for conditioning your own response to the severely disabled as well as for acknowledging their humanity.

With impaired patients who are conscious and capable of speech, address them first in a way that is consistent with their ability to respond and only after that direct your attention to a third party. When it is necessary to deal with a caregiver on specific issues beyond the patient's ability to respond, first explain to the patient that you are going to ask for information from the caregiver. Even better, ask permission and then talk to the caregiver. Amazingly, some physicians direct questions to a caregiver when a severely disabled patient has perfectly intact speech and mental competence. This is a clear form of disrespect and a form of neglect that should never happen.

Families who provide long-term care for a disabled person at home face demands that the physician can hardly imagine, let alone experience firsthand. They are uniquely sensitive to evidence of neglect or disdain for their disabled loved one. Conversely, caregivers value even small expressions of sincere concern and respect for the person in their care. This is especially so for the parents of developmentally disabled children. The same is also true of staff working in chronic care facilities. Observing the practices described above will make you a better physician and yield rewards in trust and cooperation from those who deal with these problems on a daily basis.

Religion and Spirituality

Several published studies, based on voluntary responses to questionnaires, suggest that a narrow majority of patients believe that it is at least permissible for a physician to inquire about a patient's spiritual needs (Astrow 2003). Depending largely on the strength of their religious orientations, some patients report wanting to have their physician ask about religious beliefs. But the data seems to trend toward a conclusion that the more a physician's inquiries focus on specific religious beliefs, the less patients welcome them. When surveyed, 85 percent of doctors said they ought to be aware of their patients' spiritual beliefs (Monroe et al. 2003). Paradoxically, actual inquiries about the spiritual beliefs of patients are relatively rare (Ellis 1999; Chibnall 2001).

A recent study, based on questionnaires returned by 369 outpatients at a Comprehensive Cancer Center in New York city found that while 73 percent had at least one stated spiritual need, only 6 percent were asked by any member of the medical staff about spiritual needs (Astrow 2007). Published studies that prospectively examine the practical outcome of physician efforts to attend to the spiritual needs of their patients are lacking.

At best, only tentative conclusions may be drawn from these data. Although religious faith and spirituality are highly valued by many sick people in America, medical efforts to attend to spiritual aspects of care rank very low in ratings of patient satisfaction (Koenig 2004). However, it is not at all clear exactly what, if anything, doctors should do about this. I suggest an approach based on exploring patient's needs rather than their beliefs.

Especially when the course of an illness is not going well, filled with uncertainty or beset with difficult symptoms the doctor should say something that signifies willingness to hear about spiritual needs. You could start by saying, "You have had a rough time this past week. How are you managing to cope with your problems?" The formula consists of a declarative statement that recognizes hard times, followed by an open-ended question. If a patient responds by referring to resources based on religious faith the response should be met with respectful approval. A response that seems too good to be true may be an indication of unmet needs or social isolation. Expressions of despair should not elicit blanket reassurance but sometimes are helped by validation: "you have a right to feel that way considering what you have been dealing with. You deserve a lot of credit just for showing up today. Do you have anyone else to talk to about this?" Often, simply listening works better than anything else.

The Final Visit

Even if fatal illness is rarely part of your practice you may someday find yourself at the bedside of a patient who has just died. In specialties where this is a frequent occurrence doctors can develop the ability to respond to the needs of surviving family members with skill and grace through experience. For others, confronting the unique emotional and physical environment that accompanies the moment of death can be extremely challenging. First I will discuss dealing with a death that occurs in the hospital setting.

A common situation occurs after an unsuccessful attempt at cardiopulmonary resuscitation. Typically, the scene in the immediate aftermath will be chaotic - hospital bed in awkward position, equipment and disposable supplies scattered about, staff members exiting. Hopefully, a staff member will have guided any visitors to a quiet location early in the resuscitation process. As soon as the senior physician present has given the order to suspend CPR, he or she should give firm instructions to the most senior nurse present to detain visitors until the room can be restored to some semblance of order.

As quickly as possible, tubes and monitoring equipment should be detached from the patient's body, the bed returned to normal position, a pillow replaced, bed linen straightened and any visible traces of body fluids cleaned. By all means, turn off any visible or audible monitoring devices. Nursing staff will respect any effort you make to complete these procedures quickly. Despite movie and television dramatizations to the contrary, bed covers should be folded carefully at shoulder level, not raised to cover the face. If the patient's eyes are open, efforts to close them are nearly always fruitless.

If the attending physician is not present, the senior physician among those who responded to the code blue is obligated to stand by until any visitors who were present return to the bedside. If you are not already acquainted with the family or visitors start by introducing yourself briefly. The guiding principle from that point is, keep it simple.

Elaborate expressions of sympathy are not required. The statement, "I am very sorry for your loss," will suffice. A statement in plain language along the lines of, "We tried every measure we could to restore your father's heartbeat but did not succeed," should be made. If family members or visitors ask questions, elaborate discussions of the progress of the resuscitation are not appropriate. Avoid discussion of the physiology of death. If necessary, explain how the underlying illness was too great a burden to overcome. If there is an indication of anxiety about suffering, add reassurance - for example, "He was deeply unconscious from the moment his heart stopped and completely unaware of what happened." Unless you are familiar with the details of the patient's illness, refer questions about it to the attending physician.

Family members must always be permitted to remain at the bedside as long as they wish. Ideally, your hospital will have members of the clergy on-site to take over with spiritual support and advice about disposition of the body. It is a valuable asset when a family has an established relationship with a funeral director. I have found these professionals to be responsive and helpful at all hours, day and night. If the family is completely unprepared to make decisions, it is appropriate to suggest that contacting a funeral director should be the next step. They should be assured that the hospital will care for the remains until arrangements are made. It is unkind to specifically mention the morgue at this moment.

When death occurs without the intervention of a code blue and you are called upon to make a pronouncement of death in the presence of family or visitors, the emphasis should be on dignity. Make note of the patient's name before you address the family or visitors. Move slowly. Make your gestures deliberate and formal. Even if it is obvious that the patient is dead, gently feel for a pulse at the wrist or carotid, place a hand on the patient's forehead and, only then, place a stethoscope on the patient's chest for several seconds. Turn and quietly say, "I am very sorry to say that Mr._____ has died. I am sorry for your loss." Earlier remarks about preparing the patient's body and responding to questions still apply. Keep it simple, and do not attempt to comment on details if you are not familiar with the patient's illness.

Reactions to the death of a loved one are so varied and unpredictable that any attempt to prescribe more specific responses is likely to be misleading. If you realize that you are, at this sensitive moment, representing the entire medical profession, you will not go far wrong.

CHAPTER SEVEN   FAMILY AFFAIRS

It seems obvious that family involvement can be a valuable asset to a physician caring for a sick patient. But complex family relations can also create problems. As a bride or groom marries a family as well as a spouse, a physician often assumes responsibility for the care of a family along with responsibility for the care of a sick person. Tolstoy wrote in *Anna Karenina,* "Happy families are all alike; every unhappy family is unhappy in its own way." For the physician, it is seldom immediately apparent that a patient's family is unhappy.

Even when unhappiness is apparent, unraveling which of the innumerable possible reasons applies to a particular family may be an impossible task. And even in happy families differing attitudes, goals and priorities can lead to conflict among family members or between patient and family. Of the many factors that determine the state of relations within a family only one is subject to the doctor's control. This factor, which can make family involvement an asset or a problem, is communication.

Privacy

The potential for positive or negative influence on patient care makes it worthwhile to devote some effort to maintaining good lines of communication with a patient's family. The doctor who ignores this may wind up in the middle of a conflict and the target of hard feelings on both sides. The Health Insurance Portability and Accountability Act (HIPAA) enacted in 1996 lays out strict rules for protecting the privacy of medical records.

An unintended consequence of this legislation is that physicians have become reluctant or entirely unwilling to discuss any aspect of a patient's illness with family members. HIPPA has undergone technical correction or amendment on several occasions. A document titled "HIPAA Administrative Simplification" runs to 101 pages. I hope to be excused for not trying to summarize the provisions of the Act. Those who wish to study the privacy regulations mandated by HIPAA may begin by visiting the Web site http://www.hhs.gov/ocr/hipaa/. Suffice it to say, the stated purpose and the complexity of this legislation have had a chilling effect on the ability of physicians to maintain good communication with the families of sick people.

It need not be this way. First, there has not been a wave of prosecutions of well-meaning physicians who, in an honest effort to communicate with a patient's spouse, son or daughter, break one of the privacy rules by speaking to them. Second, the proactive effort to ask a patient for a simple written permission to communicate with one or more designated family members will provide ample protection for the doctor and save a lot of frustration on both sides.

In fact, even if HIPAA had never been enacted, it would be a sensible policy to routinely ask patients who they are comfortable having you talk with and, equally important, if there is anyone you should not talk with. Finally, when one weighs the consequences of refusing to answer a simple question from a concerned spouse, son or daughter, when your best instinct tells you that you should, against the remote possibility that you might get in trouble with the Health Services Administration, the choice should be obvious.

Choosing Sides

The doctor's first duty is to his or her patient. It might seem proper, therefore, to take the position that, in all cases, the doctor's place is on the side of the patient. But what about the situation that arises when a sick person with a highly involved family is acting unreasonably or irrationally, making choices against his or her own interest? In this situation the doctor may have to respond to entreaties from a concerned relative that go like this, "You must make Dad take his medicine (stop driving, slow down, stay out of the garden, stop drinking, etc.)." Or a family has internal dissension about an important issue and leaves a sick person in the position of having to choose the opinions of some of their loved ones and reject the opinions of others? You will find yourself being asked to cast a tie-breaking vote.

What about situations in which a patient is mentally impaired or unconscious and the family is choosing a course that is contrary to instructions previously given by the patient to his or her doctor? The doctor's choice is either to violate those instructions or advocate a particular course of action against the family's wishes as a surrogate for the patient. It will immediately be apparent that there are no easy answers to these questions. They all involve a decision to take one side or the other and are appropriate for referral to an ethics committee, a resource that will be discussed in more detail below.

When competent sick people are acting in ways that jeopardize their own health, the principle of individual autonomy requires that a physician can only advise, not command. Most ethicists maintain that competent people have a right to make bad choices, and physicians recognize this at some level. When the sick person is a loved one, however, worried family members are inclined to prevail on the patient's physician to persuade the patient to make what they consider good choices. The family is also inclined to attribute more authority to the physician than he or she actually has.

In all these cases it is best not to retreat behind a recitation of your duty to the patient or the principle of individual autonomy. It is more helpful to detail for the family member what you have already said to persuade the patient to change his or her behavior. Give the family additional assurance that you will keep trying. Point out that the patient already knows what should be done and has been unable to do it. Finally, explain that presenting some sort of ultimatum to a sick person can ruin a therapeutic relationship. If the issue seems trivial to you, resist the temptation to minimize its importance. For whatever reason the family member has already decided that it matters. If possible, invite the family member to sit in on the next conversation so he or she can witness your efforts to give advice and the patient's response.

A similar situation occurs when a patient wishes to follow a course that he or she knows will inevitably cause conflict among family members and then asks the doctor to take responsibility for the choice. When confronted with this sort of request discuss the conflict with the patient before approaching the family in order to be confident that you understand the family dynamics and the patient's actual position. Ideally, when you speak to family members, the patient will be present to confirm that you are simply explaining the course your patient wants to follow. Your message should be, "This is a difficult choice. I know some of you disagree, but I am obligated to follow your loved one's wishes." It should not be, "This is what I have told him we should do." When one faction in a family is advocating a course of action that is clearly wrong there is no choice but to take the other side, explaining your opinion as tactfully as possible.

It is common for large families to be widely dispersed across the country. Especially when severe illness strikes an elderly parent, close relatives who have not been in frequent contact sometimes appear on the scene with strong ideas about the parent's care. When conflict ensues it often centers on whether to continue vigorous treatment. The shock of finding a parent profoundly impaired, after last seeing him alert and physically active, may prompt the out-of-town family members to make urgent calls for action. In contrast, family members who have witnessed a progressive decline in spite of medical attention may have decided that further efforts will be futile.

In other cases, it can seem that previously uninvolved family members are trying to make up for lost time, showing how much they care by becoming closely involved at a time of crisis and urging various new measures. The reaction of family members who have been on the scene for some time, perhaps providing major physical support, is predictable - "Where were you when we needed you?" Paradoxically, when the condition of a severely ill parent has recently deteriorated, the local faction may be subject to criticism by the out-of-town faction - "How could you have let this happen?" The atmosphere can rapidly become tense and bitter.

Imagine the consequences when an inheritance is involved. Clearly, these are worst-case scenarios, the stuff of novels rather than real life. But, if you practice long enough, you will eventually find life imitating art. The only practical policy in extreme cases is to take everything you hear with a large grain of salt, assiduously avoid statements that take one side or the other and document your decisions with great care and discretion. Last but not least, when possible, seek out the true feelings of your patient, preferably in the presence of an impartial witness.

Who Speaks When The Patient Cannot?

When severe illness renders a sick person unable to speak for himself or herself, all the situations described so far become more complex. Even mutually supportive families find the decision to abandon active treatment in favor of comfort care heartbreaking. Denial of reality, feelings of guilt, reluctance to assume responsibility for what seems to be a life-and-death decision and ambivalence characterize such situations.

It is helpful to remind loved ones who are in the grip of these emotions that they are better able than anyone else to know what choice the sick person would wish to make if he or she were able to speak. Ask whether the patient spoke about such matters in the past. Failing guidance of this kind, ask what their understanding is of how the patient would respond if we could ask what to do today. Then assert your own feeling that, in a sense, when a loving next of kin acts on the basis of a sincere belief in what a patient would want, it is as close as we can get to the patient's own decision.

In the face of an intense family dispute over what to do it is not possible for a doctor to act as surrogate for a patient without provoking more conflict or becoming the target of anger. There is no language so persuasive and carefully crafted that it can guarantee a satisfactory resolution. But there are two resources that are readily available and should always be called upon when the patient's family is not united.

The first is a preventive measure. Before a crisis develops, do not fail to ask your patients, including healthy patients, to execute a document called an Advance Directive or Durable Power of Attorney for Health Care depending on your location. Although an attorney can prepare this document, these forms are also available in nearly every hospital in the United States as well as on the Internet. Every state has its own version of the form. The Web site of the National Hospice and Palliative Care Organization offers all fifty versions without charge (www.caringinfo.org/stateaddownload).

Remember that a Durable Power of Attorney designates a surrogate decision maker but may not detail the individual's preferences about specific medical decisions. Furthermore, if this document merely chooses yes or no for cardiopulmonary resuscitation (CPR) in the event of sudden death, it will leave the most difficult situations that can occur without any guidance (Tulsky 2005). I will address the specific question of how to discuss the choice for or against CPR in the next chapter.

The second resource is the hospital ethics committee. These groups vary widely in composition and procedures so it is worthwhile to become acquainted with the committee at your hospital in advance of need. In the best examples clergy and nursing staff are represented in addition to medical staff. Although the findings of an ethics committee are only advisory in most institutions, the process, when it works well, can be invaluable in defusing tension and conflict. The process takes the focus off the treating doctor as the exclusive decision maker. Family members are invited to participate and can express their views in a structured, neutral setting that gives them confidence that they are being taken seriously. Even when there is no "right answer" the process can lead to resolution.

Too Much or Too Little

For many families, the most trying decisions involve the use of supportive measures when meaningful recovery is an unrealistic hope. Such situations can occur after a long downhill course of illness or suddenly after traumatic injury or sudden collapse. These end-of-life decisions center on which, if any, of the full arsenal of critical care, including constant monitoring of metabolic balance and cardiac function, mechanical ventilation and blood pressure support, are appropriate.

Although these measures can prolong life, near the end of a long terminal illness or after irreversible brain injury they cannot be expected to restore an individual to a functional state.

An Advance Directive that specifies a patient's wishes for or against these forms of supportive therapy, as well as CPR, can be enormously helpful to resolve the conflicts felt by families who confront terminal illness in a loved one.

In many of the standard forms currently available, the patient's directions about when to withhold these specific supportive measures have to be written out by hand rather than simply by checking boxes. Recognizing that patients' wishes in respect to life-sustaining treatment were not being consistently honored in spite of the availability of Advance Directives, Oregon introduced a comprehensive program to improve care and an improved document in 1991 (Meyers et al. 2004). This form (Patient Orders for Life Sustaining Treatment) has recently been adopted in California as well and should serve as a model for other jurisdictions.

The use of artificial means to provide fluids and nutrition falls into a different category that should also be addressed in an Advance Directive. Food and water are so basic to survival that some individuals will equate withholding either with "letting someone die" or, worse, killing. Others will argue that, since they might relieve thirst or hunger, these forms of therapy are appropriate even when they only prolong the process of dying. When a patient is unconscious it is understandable that loved ones who witness dehydration, wasting, respiratory distress or fever will be profoundly distressed. They may also project their own distress on the sick person even when that person is deeply unconscious.

Acceptance of a decision to forgo burdensome or heroic measures to sustain life ultimately depends on acceptance of the validity of the following concept: the underlying illness is the process that will end the patient's life, not starvation, thirst, medicine to relieve suffering or cessation of mechanical ventilation. It will not always be possible to gain this acceptance. But cautious explanation of several facts should be included in any discussion.

Distressed family members should be given gentle reassurance that thirst can be alleviated by careful attention to oral hygiene, that the severe anorexia of advanced cancer eliminates hunger, and that morphine in judicious doses is effective in relieving dyspnea and will not cause breathing to stop. In giving these explanations, the emphasis must be on the patient's comfort. After a reminder of the reality that the illness has progressed in spite of best efforts to reverse its course explain that pressors, transfusions, respirators and so on merely support physiologic processes without hope of changing that reality.

This is emphatically not the time or place for discussions of appropriate use of scarce resources or the expense of extraordinary measures. If these reassurances are to be credible, they must be backed up by exemplary nursing care. It should be noted that the physician's attitudes set the tone for attaining the best nursing care. This is time when careful attention to hygiene and appearances is of vital importance.

Finally, in all end-of-life situations, physician presence becomes a crucial element in maintaining goodwill. Frequent visits to the bedside communicate sincere caring more than do words. Avoid by word or action anything that might suggest withdrawal on your part. Be tolerant of families that seem to "camp out" at the bedside. Conversely, when a family member who has previously been a primary caregiver suddenly becomes hard to find, do not assume the individual has become uninvolved. Firsthand observation of the dying process may have simply become too painful for the family member to bear. In such cases, try to stay in contact by phone.

CHAPTER EIGHT   LOADED WORDS

Certain phrases and subjects seem fated to provoke misunderstanding or hard feelings with a frequency that warrants special attention. Some of these are discussed below, in no particular order.

Promises

"I will get back to you." "I will send the report." "I will call in the prescription." "I will talk to the consultant." Such casual statements made so often and so easily in the course of a busy day represent a commitment. In some settings - conversations between friends, business calls - failure to keep a commitment may assume the character of a broken promise and cause serious irritation. But in the setting of illness the reaction may be far out of proportion to the practical significance of what was promised.

The reasons for this should be obvious. First, the two parties are unequal. The doctor, correctly or not, is perceived as having the power to make things happen. The patient is the one with the disease. To be sure, doctors care about outcomes, but they cannot match the concern a sick person feels about his or her illness. In addition the doctor has many competing priorities and is working hard to respond appropriately to other patients with conflicting demands. The sick have only their own symptoms, fears and uncertainty to occupy their thoughts.

Finally, the patient usually has to penetrate a screen of intermediaries (phone mail, receptionists, nurses, etc.) just to reach the doctor. While people are usually remarkably tolerant of small failures on the part of their doctors the effect of repeated small failures is cumulative. When a patient's tolerance is finally exhausted a formerly good relationship may be badly damaged.

What is to be done to prevent this? First, be aware that what seems to be a matter of minor significance to you may be seen as a vital matter by your patient. For the doctor a normal test result means no further action is needed. To a worried patient it may represent relief from his or her worst fear.

Second, the natural impulse for physicians, anxious to serve their patients and perhaps overconfident, is to promise more than they can deliver. But it is unrealistic to expect that you will reach perfection or even close to perfection in keeping every commitment you make. To avoid this pitfall, exercise discipline in not making commitments you know in advance will be difficult to keep. Thus, "I'll get back to you about the results" is a safer promise than "I'll call you this evening." No patient will ever criticize a call that comes earlier than expected. And a call that comes when it was not requested or promised will be treated as a generous gift. Train your staff to alert you to at the end of the day to loose ends that result from these promises.

Third, for commitments that depend on the action of a third party - another medical office, a laboratory or a pharmacy - be sure to make it clear to the patient that you are counting on the third party to complete the action and, where possible, limit your collaboration to third parties you can count on. When feasible, as a fail-safe, give the patient clear instructions about what to do in case the intended action does not happen instead of relying solely on your own memory.

Fourth, remember that everything your staff does and says reflects on your own attitudes and standards. The doctor's staff, whether because of a well-meaning desire to "protect" the doctor from distractions or because of inefficiency or incompetence can easily present a picture far from what the doctor wishes. Regular meetings with the staff, objective measures of performance and explicit statements that make your attitudes and standards clear are required to maintain the image you wish to present to your patients.

Criticism

Criticism rarely has a place in a physician's communication with patients. Unless there is an actual threat to a patient's safety, criticism of other doctors or medical facilities serves no useful purpose. The criticism does not have to be overt to be damaging. Perhaps a patient comes to see you with results of a test ordered by another doctor that you feel was inappropriate. Thinking out loud, you say, "I can't imagine why he ordered that test." This statement does not contribute in any way to the patient's recovery. It does, however, have the potential to harm what might be a valuable preexisting relationship with the other doctor. Furthermore, if it later turns out that there was a valid indication for the test that you were unaware of, you may have given the patient reason to wonder about your own competence.

Avoiding criticism of other professionals has nothing to do with a code of silence. Refraining from casual criticism does not interfere with the obligation of the profession to police itself. There are plenty of avenues for bringing professional incompetence or misbehavior to the attention of hospital staff officers, medical organizations or regulatory authorities. But a patient has no way to respond to critical statements except with a loss of confidence in the profession as a whole. When your own findings require you to contradict the diagnosis or therapeutic plan of a previous physician, find a way to tactfully explain that you now have the advantage of additional data that leads you to a different conclusion from the one reached earlier. A sick person confronting a major change in diagnosis or treatment has enough to worry about without the added burden of recriminations or a loss of trust in another doctor.

It is important to exercise care in listening to criticism as well. Sick people will sometimes relate an unpleasant experience or unsatisfactory result and ask for your opinion. Although the story might sound compelling, venturing a judgment on the validity of the complaint or the extent of harm done is unwise. Even cautious validation of criticism of others may lead to the assumption that you will support some future claim for compensation.

Unless you are certain that you have all the facts you are not in a position to foster that assumption. If it turns out, in the light of additional facts, that you have to retract your support for that claim you may face disappointment and anger. On the other hand, when the patient's complaint seems clearly unjustified it is probably better to express a neutral observation such as, "I'm sorry that happened." than to leap to the defense of the other party. Again, you do not have all the facts, and defending a third party puts you in the position of an advocate for an entity with which your patient is already upset and sets the stage for an adversarial relationship between you and your patient.

Considering the complexity of medical practice today and the volume of media attention to medical errors, it is surprisingly uncommon for sick people to express criticism directly to their doctors. On the contrary, it seems that patients are more likely to try to please their doctors, even when they may not be totally happy with the way things are going. Because it is impossible to respond to critical feelings that are hidden, perhaps we would all be better off if patients complained to us more often. Possibly because criticism from patients is relatively rare, doctors may be caught off guard when they experience it. In any case, unlike politicians and other public figures, physicians seldom become adept at responding to complaints about their own performance.

When the criticism concerns your office policies or administrative matters, simply giving the complaint a fair hearing without interruptions or excuses may be sufficient. When you are confident that the commitment can be kept, making a pledge to correct the problem is even better. When your own performance is the subject of the complaint, an acknowledgment that the complaint is valid and a simple apology, again without excuses, is best even if you must overcome a sense of injustice.

Although you may feel the complaint is less than fair, remember that for most people perception is reality. A generous response without rationalizations from someone in authority (such as a physician) will go a long way toward defusing a potentially tense exchange. Medical errors that involve actual or potential harm are another matter altogether. They pose an especially complex set of issues that I will address at the end of this chapter.

Why Didn't You Call Me?

Sick people, like anyone else, can be very exasperating at times. They may ignore instructions, forget to take medicine, fail to report symptoms, lose track of instructions, miss appointments and commit a variety of other breaches of their responsibility to "be a good patient." When you are at the end of a long, hard day, taking a phone call at 4:45 on a Friday afternoon or have just had a frustrating interaction with the health care bureaucracy, it is difficult to resist pointing out one of these breaches to a sick person and explaining why the breach has made your job more difficult.

No matter how satisfying one of these lectures might be in the short term you should resist the impulse to give one. For a number of reasons, sick people find it difficult to be objective. Above all, they do not enjoy being scolded. For one thing, most of us acquire a sense of entitlement when we are ill. And the sicker we are, the more strongly we believe that we deserve immunity from criticism.

For the sick, criticism, however well justified, and especially if delivered by the individual to whom they have turned for help, is especially unwelcome. There is also a good chance that the offending patient is well aware that he has erred and is either apologetic or expects to be chastised. The doctor has a perfect opportunity to demonstrate a caring attitude by saying something positive instead of something negative.

Rather than "Why didn't you call me sooner?" Try saying, "I'm so glad you called me, now we can get to work on this problem." Rather than "Why didn't you take your medicine?" try, "I know it is hard to keep track of all these pills, but they will work better if you can find a way to take them regularly. Let's figure out a way to help you do that." Your patient will first be surprised and then immensely relieved and grateful for your generous response. He or she might even be motivated to do better in the future rather than feel resentful or abused.

Brave Talk and Lavish Praise

"I am not afraid of death." "I know I'm going to recover." "Give it to me straight, no matter how bad it is." "You are the best doctor I have ever met." "You saved my life." Statements of great optimism and courage or flattery of this sort may be perfectly sincere. But sometimes they arise completely out of context. Or something about the statement causes you to sense that something is not quite right. The words seem discordant with the actual situation. Perhaps death is by no means imminent, no earthshaking news is in the offing or you have done nothing to deserve special praise. In such cases doctors should look beyond the words themselves and ask what has prompted such bold or flattering talk.

Consider the possibility that what you are hearing serves a special purpose for the speaker. Expressions of bravado could

144

be the product of denial - a mechanism for protecting oneself from the impact of a reality that is too difficult to bear. It is also possible that maintaining a brave front is a sign that the patient has embraced the popular notion that "a good attitude" is required in order for treatment to be effective. Problems arise when the brave front is serving one of these purposes. If it is only hiding anxiety such talk will prevent intervention with more useful forms of emotional support. Brave talk is a poor defense because it always fails when the sick need it most, at 4:00 A.M., alone with their own thoughts and facing an uncertain future. And the belief that one must have a positive attitude if one is to get well becomes damaging if it leads to the corollary belief that experiencing fear or pessimism causes failure.

For the physician, an effective strategy for managing the flawed emotional defense of brave talk is to start by validating the patient's expression of optimism and bravery in spite of your instinct that it may be a cover for underlying insecurity. Instead of questioning the validity of the statement, say, "I admire your confidence and good spirit." Then follow up immediately with a potentially more durable form of support. For example, "It is most helpful to have courage, but remember, we all have moments of weakness. It is normal to have negative thoughts, worries or even fear once in a while. But *that has no bad effect at all on your illness.*" If you think it is necessary, drive this point home: "Your illness doesn't know what you are thinking."

In framing your support this way, you may succeed in immunizing your patient from fearing that the pessimistic thought that inevitably enters her mind can harm her chance for recovery - without directly challenging the sincerity of the statement.

Exaggerated praise of one's doctor sometimes serves to reinforce a kind of magical thinking on the part of the patient. Such thinking rests on the idea, "If my doctor is as good as I say he or she is, I can't fail to get better." The problem with this idea is that by staking success on an unrealistic picture of what the doctor can accomplish, a sick person may be setting himself up for a complete loss of confidence or feelings of betrayal when things do not go as well as hoped.

Especially in the case of lavish praise that is clearly out of proportion to reality, it is important to resist being seduced by the pleasant experience. A statement by the doctor that conveys the following ideas can be useful. "Thank you for the kind words. I will continue to work hard for you. But remember, you and I have lots of powerful assets helping us in this effort: talented specialists, powerful new drugs, ongoing research and your wonderful family. I am doing what any doctor would do. Lots of other doctors could take over if I were not here. *Good medicine doesn't require superstars.*"

Code Blue and DNR

Many hospitals at the time of admission ask patients to execute a document expressing their preference for or against cardiopulmonary resuscitation. Thanks to the dramatic impact of television and film depictions of cardiopulmonary resuscitation, virtually all American adults believe they have a clear idea of what this procedure entails. However, few laypeople are aware that in published studies of CPR in hospitalized patients the rate of survival to hospital discharge ranges from 13 percent to 37 percent, with most results at the lower end of this range (Schneider et al. 1993).

Even fewer laypeople understand the difference between the prospects for a person who sustains a cardiac arrest on the operating table and one who is found without a pulse in bed in a hospital room. The former has a fair chance of leaving the hospital on his feet. The latter rarely does. Virtually all patients who survive CPR spend at least some time in the ICU on a mechanical ventilator after a pulse is restored.

Only the rare layperson has ever witnessed a real code blue, which is usually more chaotic and far less tidy than the movie or television version. It is important to use language that is both simple and precise in discussions about CPR with a sick person. By "simple" I mean language that is free of jargon, acronyms and technical terms.

For example, a discussion could start like this: "In this hospital the standard emergency procedure when someone's heart suddenly stops is to try to make it start again. This is done by thumping the breastbone and giving an electric shock across the chest. Unless a strong pulse returns immediately, it is then necessary to press firmly and rhythmically on the breastbone to pump blood out of the heart. It is also necessary to place a plastic tube from the mouth to the windpipe in order to blow air into the lungs. While this is happening, powerful drugs are given through an IV. When this procedure is done for someone who was in generally good health before his or her heart stopped, it is often successful.

"When a person had other severe medical problems that were present before his or her heart stopped, the procedure is usually not successful. Even when the heartbeat is restored, a person who was very sick before his or her heart stopped seldom returns to good health. If the heartbeat is restored the patient almost always has to spend time in the intensive care unit. If your heart stops while you are in the hospital this procedure will be done for you unless you give us instructions not to do the procedure. After considering all this, how do you feel about whether to have this procedure in case your heart suddenly stops?"

This may sound terribly long-winded and a bit stilted. But if shortcuts and euphemisms are used, if the prospect of failure or a long, futile stay in the ICU is not mentioned, a full and fair picture has not been given. Speaking slowly with pauses at strategic points, it takes only about two minutes to recite the language given above. If there are questions, and they should be encouraged, answering them should be part of the process and might only take a few minutes more. If it leads to a thoughtful, informed decision about choosing the instruction, Do Not Resuscitate, the time is well spent.

There Is Nothing Wrong

People often report symptoms that, although very real to them, defy explanation in a disease-based medical model. The problem for the physician is to distinguish a symptom that is inexplicable but truly innocent from one that is inexplicable only because the disease responsible for the symptom is hidden. Some can be identified as actual physical sensations or alterations in performance related to normal physiology but misperceived as signs of disease by an anxious person.

In such cases it would be technically correct to say that nothing is wrong, at least insofar as "something wrong" is taken as synonymous with the presence of a disease. Communicating this conclusion effectively requires the use of carefully chosen language, and none of that language includes the words, "There is nothing wrong" (McDonald 1996). When you are convinced that a symptom is the result of a normal physiologic process your task is first to acknowledge that the symptom is real and then to describe the physiology in terms that are understandable and persuasive.

Success in explaining this phenomenon requires convincing a worried patient, perhaps one with a constitutionally low threshold for anxiety about health, that his or her sensations are harmless. In this context, *normal* is a word fraught with significance. Casually tossing off the phrase, "That's normal," without some objective data to support the conclusion usually provokes more doubt and anxiety than existed before (McDonald et al. 1996). In fact, without some minimum of essential objective data, this statement is unjustified.

A better strategy is to list the results that support the conclusion and then explain what and how ordinary physiologic processes can produce such feelings. Even better is when simple advice about diet, exercise or behavior can offer the hope that the sensations will diminish or go away. Finally, it is wise to include a fail-safe provision that invites the patient to return for more evaluation if the symptom worsens or changes. Again, if you can't conceive of a physiologic explanation, question the validity of your conclusion.

When a patient's emotional profile suggests that he or she might experience increased anxiety rather than relief after a normal test result, it is useful to explain in advance the meaning of a normal test result before the test is performed. A study of patients with chest pain showed this to be the case (Petrie et al. 2007).

Early in the course of a medical evaluation and especially when you cannot make a convincing case for an innocent cause, it is may be prudent to follow the principle of assuming that symptoms are due to illness until proven otherwise. But when an evaluation has proceeded through the process of history taking, physical examination, laboratory tests and radiology studies without revealing any evidence of disease it will become necessary to decide when to stop looking for one. In fact, continuing diagnostic procedures past a point of diminishing returns may actually increase anxiety in some patients (Page and Wessely 2003).

In addition to the practical issues of time, effort and expense, increasingly complex test procedures involve discomfort and the risk of harmful complications. Once you make a decision to stop further investigation you must give an explanation that fulfills at least two criteria. It must be clear enough to be understood, and it must also be convincing enough to satisfy a worried patient.

One useful framework for this explanation consists of reviewing with the patient the doctor's process of differential diagnosis, that is, formulating a list of the possible explanations for the symptom and then applying the results of observations and tests to rule out each of the possible diagnoses. As mentioned earlier in the discussion of reassurance, it is wise to refer to general body systems or categories of illness rather than name potentially frightening specific diagnoses. This approach has the added advantage that verbally reviewing your reasoning forces you to critically appraise your thinking. If the conclusion is not completely convincing for the doctor it certainly won't be for the patient, and perhaps further studies should be done.

If the diagnostic evaluation is done well and leads to completely negative results the possibility of illness will be very low. The problem is that the possibility of hidden illness is rarely, if ever, reduced to zero, and most people know this intuitively. Therefore, one more message must be communicated. Start by acknowledging explicitly what the patient already understands: in medicine it is hard to prove a negative, and one can rarely be 100 percent sure there is no illness.

Continue by stating that, because you know this, you will remain vigilant for further developments. Finally, and most important, explain that conditions that present an immediate danger have been completely ruled out. In other words, even though you are only be 99 percent sure that the symptoms are innocent, you can be 100 percent sure that *no harm will come* from deferring more complex, potentially risky studies until more evidence becomes available or the symptoms go away by themselves. The latter event is, by far, the most common resolution when these conditions are met.

There Is Nothing Left To Do

In a way "There is nothing left to do" is the mirror image of "There is nothing wrong." In this case, everyone knows something is dreadfully wrong but there are no good treatment options. The approach to talking about this subject was addressed in chapter 4 under the heading, "Honesty versus Hope." But here I wish to emphasize what not to say. It may seem counterintuitive to remain positive while acknowledging that an illness is incurable, but it is never more important to try. Again, consider patients who turned to unconventional therapy because they were previously told, "There is nothing else to do."

However sincere and well meaning it may have been, this statement, at best, can only be partially true. That intelligent people turn to unconventional therapy proves there is always *something* to do: not necessarily something useful, easy or even sensible but always something. The physician's responsibility is to identify measures that are both useful and possible and make a convincing case for the idea that they are worthwhile. A much more difficult task is to be persuasive when it is in the best interest of a patient to show that a proposed treatment is not worthwhile.

Recall that a seriously ill person has probably been the recipient of volumes of material heralding the promise of both legitimate and dubious treatments and that severe illness leads most of us toward a willing suspension of disbelief. When confronted with a statement that there is nothing more to do individuals in the grip of this psychology will go to great lengths, often to their own great disadvantage, to find a source that will contradict that statement. Having witnessed the sad consequences of fruitless searches for miracle cures, I believe doctors have an obligation to do what they can to spare patients and families the burdens of one of these terminal quests. The first step is to vow never to say, "There is nothing more to do."

Instead consider temporizing. For example, "We should concentrate on building up your strength before going on to a very taxing form of therapy." Or, "We should wait until you have recovered from this complication before introducing a new, potentially toxic treatment." A physician can emphasize treating a secondary but potentially treatable aspect of the patient's illness rather than a direct assault on the underlying diagnosis. This not only puts off an irrevocable commitment to burdensome treatment that is bound to fail but also offers the possibility of some small success in relieving symptoms, sorely needed after a series of failures. Do these statements obscure harsh reality? Sure they do. But the sick person may already be in a state of vigorous denial. In a situation like this I believe it is ethical for a caring physician to become the patient's accomplice in denial.

This strategy, in essence playing for time, differs from saying, in effect, "Enjoy what time you have left." Such casual advice ignores the fact that the presence of life-threatening illness is usually accompanied by physical or emotional limitations that foreclose the possibility of travel or enjoyment of pursuits that were once highly valued. I was frequently surprised at how often people with fatal illness fantasize about travel in ways that seemed quite unrealistic. I suspect that, when faced with one's own mortality, travel can symbolize escape from a painful reality. Thinking about it provides a form of comfort even though the plans are rarely acted on. For this reason, it is not helpful or necessary to point out why the plans are impractical.

In essence, remaining positive in dire situations requires a doctor to be diligent in trying to identify an achievable therapeutic goal, however small, and to approach it with the same sincerity and attention to detail that would be devoted to a dramatic surgical procedure or complex drug regimen. This approach sometimes strains the limits of credibility, but, accompanied by a large dose of compassion, it may be enough.

Psyche and Soma

It has been known since ancient times that the mind influences one's physical state. Our understanding of the mechanisms for this relationship is incomplete, as illustrated by the evolution of related diagnostic language. The label "hypochondriasis" arose when dysfunction of the upper abdominal organs was thought to be responsible for illness in the absence of obvious disease. When Freudian analysis came to dominate the practice of psychiatry, conditions once understood as due to the influence of the uterus (hysteria) acquired a scientific gloss as psychosomatic illness. Later in the twentieth century the diagnosis "functional illness" was considered more appropriate. The current version of the *Diagnostic and Statistical*

Manual of Mental Disorders (DSM IV) defines "Somatoform" disorders with six subcategories including hypochondriasis and conversion reactions (Noyes et al. 2008; Stone et al. 2005). I still prefer the term "functional illness" because it is simple, refers to the contribution of normal physiology and sounds less like jargon. Whatever language we use to describe these conditions, a significant percentage of patients seen in general practice and internal medicine fall into this category.

Their symptoms commonly include fatigue and weakness, soft tissue, joint and muscle pain, evanescent rashes, sensitivity to a variety of external agents, disordered gastrointestinal motility, headache and cognitive difficulty. When caring for a patient for whom no other diagnosis seems to explain the clinical picture it is prudent to recall that peptic ulcer disease was considered a prime example of psychosomatic illness until the discovery of *Helicobacter pylori* twenty years ago. Susan Sontag, in her excellent book, *Illness As Metaphor*, explored the psychological profiles thought to underlie tuberculosis, syphilis and cancer until the actual causes were identified (Sontag 1978). Polymyalgia rheumatica, a serious and clearly organic illness, can easily masquerade as functional illness. We would do well to remain humble when approaching puzzling conditions such as chronic fatigue syndrome, fibromyalgia, irritable bowel syndrome and Gulf War syndrome, any one of which may someday come to be understood in purely biologic terms (Maes 2009).

As discussed above, the perception of harmless but real physical sensations often leads healthy individuals to become convinced that they are ill. Emotional sensitivity, an unusually reactive autonomic nervous system or a variety of difficult life situations can all predispose to this phenomenon. When talking about it with a patient, it is all too easy to provoke feelings of defensiveness, shame or resentment. If you have previously displayed skepticism or impatience in the course of an evaluation or, worse, hinted that the patient is imagining symptoms, seeking attention or malingering, hope of reaching a good outcome is probably already lost.

Aside from being time consuming, functional illness tends to persist, resisting even determined efforts to help. In the absence of a clear remedy, the patient's need for attention tends to increase the number and urgency of symptoms over time. The lack of adequate compensation for the time required to properly manage the problem presents considerable practical difficulty. These problems, when combined with frustration produced by the inability to help someone in distress, create a fertile setting for impatience on the part of a physician - even resentment. Exemplary patience and a generous spirit are clearly required if one is to resist the tendency to take shortcuts in management.

Such shortcuts employ what could charitably be called diversionary tactics. One such tactic is to choose a non-illness with a pseudoscientific label (trace element deficiency, low blood pressure, hiatus hernia, hypoglycemia, chronic yeast infection, to name a few) to explain the symptoms. This might lend legitimacy and help to make sense of the problem (Page and Wessely 2003). Unfortunately, this tactic is not only deceptive, but inevitably leads to wasteful, fruitless and ultimately harmful therapies. Or one can resort to prescribing various pills, injections or procedures whose real purpose is to get the patient out of the office (vitamins, tranquilizers, supplements) or to seize upon some lab result or anatomic finding of dubious significance to give an objective explanation for the symptom. None of these shortcuts lead to lasting benefit, and they are all ethically questionable. Antidepressants may have a place in treatment but should never be used to substitute for effective communication and a therapeutic doctor-patient relationship (Burton 2003)

It is far wiser to spend the time and energy required to make a sympathetic but forthright presentation of the problem. The presentation must include a promise to continue to work with the patient to seek safe, practical measures to lessen the impact of the symptoms. Instead of minimizing the importance of the symptom, the doctor should validate the patient's distress with an acknowledgment that he or she is having a rough time. An explanation of functional illness must always include firm reassurance that you do not think the patient is imagining symptoms or somehow responsible for his or her distress. When appropriate, an explanation of how the autonomic nervous system produces the distress will help. A reminder that many productive, admirable people experience similar problems will reinforce the point that the patient is not to blame for the problem.

Because these patients have often been rejected in the past and fear abandonment, it is critically important to state that you will continue to be available, especially if you plan to make a referral to another physician. The diagnosis of functional illness must not be seen as signing off on the care of the patient. Merely being confident that you are willing to stay involved and listen to complaints in a nonjudgmental way is often enough to lessen to intensity of a patient's concern. In rare cases, when complaints are bizarre or there is evidence of a thought disorder, psychiatric referral will be appropriate.

"I Made a Mistake"

It will happen eventually. Through oversight, misjudgment, lack of information, lapse of memory, carelessness or some other, all too human weakness, you will almost certainly make a mistake. In medicine the possibilities for error are endless. An action that should have been taken is taken too late or not at all. A decision or piece of advice turns out to be wrong. When an error does not lead to any adverse consequences it may be looked upon as a near miss, an opportunity to change procedures to ensure that it never happens again. When an error causes harm to a patient the stakes are much higher. Obviously, the first priority when this happens is to do what can be done to mitigate the damage. This discussion is limited to what to say about it.

The principles of truth telling, informed consent and justice provide a strong ethical basis for disclosing medical errors. This is not to say it is easy to do so. Doctors dread the possibility of sanction by employers or hospital staff organizations. And, given the current state of medical liability litigation in the United States, a doctor's response cannot help but be influenced by fear of a malpractice claim when a mistake causes harm. Some legal defense experts have advised against ever acknowledging a mistake. Physicians who practice in high-risk specialties justly feel that they deserve credit for taking the responsibility to care for desperately sick people.

When things go wrong, most of us are already experiencing intense remorse and self-criticism when the question of disclosure arises. When a mistake is discovered doctors feel hurt, vulnerable or defensive, or all three. A patient who has been the subject of a medical error, whether major or minor, is a living reminder of our failure to achieve or, perhaps, even come close to perfection. These factors help to explain why patients relate the experience of sensing that their doctors were avoiding them or seemed distant after an error has occurred.

For all these reasons doctors find it difficult to talk to patients about errors. Yet disclosure in one form or another nearly always must eventually be faced. Surveys and focus groups indicate that the elements of disclosure most important to patients include an explicit statement that an error occurred, an explanation of the nature of the error and its effect on the patient, why it happened, what corrective action will be taken and, perhaps most important, an apology.

A number of published reports document that while a majority of doctors in the United States support the concept of disclosing errors, in actual practice they do so on their own in less than half of instances (Mazor et al. 2004; Gallagher et al. 2006; Kaldjian et al. 2007). Having no legal expertise, I hesitate to make specific recommendations for dealing with the risk of professional liability, with one exception. The exception is to note that any action or statement that could possibly be construed as a cover-up will be fatal to an effective defense.

Beyond this, it is worth noting that the traditional position against ever acknowledging a medical error, let alone making an apology, has recently given way to a more balanced view (Kraman and Hamm 1999; Kachalia et al. 2003; Pelt and Faldmo 2008). A number of anecdotal reports suggest that a candid explanation of what went wrong, forthright acceptance of responsibility and an expression of sincere regret does not increase the probability of malpractice litigation and may actually lessen it. Hard evidence for this view, however, is still lacking (Mazor et al. 2004; Berlin 2006).

Five states have enacted laws mandating disclosure of medical errors (Nevada, Florida, New Jersey, Pennsylvania,

Vermont). And in recent years twenty-nine states have enacted legislation that prevents an expression of regret from being introduced as evidence of liability in a civil action. In other words, the legislation provides immunity for saying you are sorry. Federal legislation toward this end has also been introduced. You may check the status of the law in your state on the Web site of the Sorry Works Coalition (www.sorryworks.net), an organization of physicians, lawyers and risk managers.

If you are faced with the prospect of disclosing a medical error be sure to confront your own emotional response to the situation first. It is too much to expect to be able to speak sympathetically, objectively and accurately if you are still coping with feelings of guilt, remorse or insecurity. Guard against statements that sound defensive and any temptation to rationalize or excuse the mistake. By all means, seek the help of a hospital risk manager or representative of your malpractice insurance carrier before embarking on this arduous task.

CHAPTER NINE   TAKING CARE OF YOURSELF

D octors are human, susceptible to all the same limitations and frailties that beset everyone else. In spite of this, cultures and societies throughout history have usually accorded special status to healers, granting them special access, privilege and prestige. In return for this special status, laypeople expect a high level of performance on the part of their doctors, denying the human limitations and frailties that doctors share with everyone else. In short, doctors are held to a higher standard. This implicit social contract requires that physicians recognize their own physical and emotional limits and learn to reconcile these limits with the demands placed upon them by their profession. This chapter focuses on some of the most common and important of these limitations.

Fatigue

I have presented a picture of doctor-patient communication that is more than a little idealized. While grousing about the shortage of time that characterizes medical practice in America today I have urged ways of talking to the sick that can only be expected to consume even more time. It is fair to ask if I have any advice about how to resolve this contradiction. A first step is to make an objective analysis of your workload, goals and priorities. If you are part of an organization that denies you the authority to determine your own schedule or limit time at work, my advice may be of limited value.

However, if you do have a measure of control over your workload, ask yourself if you habitually overbook your office schedule, accept more new patients than you have time to manage or promise more than you can deliver. If economic considerations or a critical demand for your services leave you no choice but to continue, your options are also limited.

If neither of these conditions applies, you at least have the freedom to choose to modify your workload. If your daily schedule of office and hospital visits, phone calls, correspondence and travel does not realistically leave sufficient time for listening and talking to patients, go one step further and examine the balance in your life between work, rest, personal affairs and family activity.

If you do not have sufficient time to listen to patients and this balance is excessively weighted toward work, having reflected on your goals and priorities, ask whether changing your schedule to accommodate improved communication with patients, rest and a full personal life is important enough to accept the sacrifices necessary to make the changes. The sacrifices may be more than economic.

Often the most difficult sacrifice involves the recognition that you cannot do everything for everyone. Nearly everyone who embarks on a career in medicine starts out with altruistic feelings. Traditions of service and selfless devotion to care are instilled during medical school and postgraduate training as they were in earlier times. However, learning how to say no and set limits are skills that are generally not taught.

These skills may even be subtly discouraged, especially during internship and residency where the can-do spirit and heroic deeds of endurance are entrenched in the mythology of medical training. I vividly recall how working to exhaustion was considered a badge of honor during my own internship. One fellow resident was famously remembered for an unusual clinical phenomenon known as the "Z-lines of Farley" (a pseudonym). The Z-lines of Farley consisted of distinctive marks - diagonal streaks of ink - on Dr. Farley's hospital charts, created when he fell sound asleep over his notes. Work habits like these tend to carry over to professional life after training.

Admirable as the ethic of hard work may be, it should not be considered self-indulgent for a doctor to structure his or her schedule to permit high-quality communication with patients. One need not feel guilty about limiting one's patient load to what can reasonably be managed while maintaining high standards of performance. In fact, working while exhausted is just as likely to lead to poor performance and medical errors as are poor training or inadequate preparation. Following a series of well-publicized and tragic medical errors committed by physicians in training, organizations that accredit medical training programs attempted to mandate limits for the on-duty hours of interns and residents. The fact that these limits are still widely ignored indicates how powerful the tradition of overwork in medicine remains.

The practice of systematic overbooking in medical offices is one consequence of this tradition. Declining reimbursement for physician services - especially cognitive services - and organizational demands for increased productivity also contribute to this practice. Chronic overwork and fatigue follow. Patient communication suffers first, but so do the doctor's family life and, sometimes, the quality of his or her performance. Nobody wins when the doctor is exhausted. Most physicians learn how to deal with occasional episodes of overwork. Those who cannot function with temporarily punishing workloads and short periods without sleep can make a wise career choice for a field that permits more control over their time.

But the ability to tolerate overwork is a mixed blessing at best. Even when we manage to compensate for the purely physical effects of fatigue, the mental and emotional effects inevitably continue to accumulate. By the time others recognize the hallmarks of prolonged fatigue such as irritability, inattention and loss of humor, the affected physician is already at risk for poor performance, burnout and substance abuse. It is only prudent for doctors to take a personal inventory for these symptoms periodically. This sort of honest self-appraisal is the key step in managing fatigue. When such symptoms are present you owe it to yourself and your patients to make adjustments in your practice.

Even if it is not possible to make time for a day off or a full-scale vacation, simply setting aside a brief but regular break in the course of the workday can be surprisingly effective. Industry has long since recognized the value of break time for improving productivity. A work break for the physician of only a few minutes can take the form of listening to music, picking up a book, meditation or any other activity that interrupts the stress of medical practice. For some, although it requires more time, vigorous physical activity will be more useful than cognitive activity.

It is remarkable how even a short spell of any kind of diversion can interrupt the vicious cycle of working faster in order to make time to work harder still. Cultivating a craft, hobby, art form or athletic activity after a day's work is also an invaluable way to regain some perspective on the world away from medicine. No single technique works for everyone, but you owe it to yourself to identify one that works for you and then use it.

Depression

The high expectations the sick and their loved ones hold for what modern medicine can accomplish seem to increase more rapidly than actual advances in technology and science. By the same token, doctors, having completed competitive and strenuous preparation for their careers in medicine and highly motivated to do well, set ambitious goals for themselves. But medical outcomes will always be associated with an element of random chance.

Diagnostic tests that are reliable under normal circumstances are sometimes misleading. Therapy that is nearly always effective will sometimes fail for no apparent reason. Some diagnoses continue to defy our best efforts to achieve a cure in spite of the advance of medical science. Beyond the role of chance in determining medical outcomes, economic pressures, fatigue and the pernicious influence of various third parties (insurers, government, managed care organizations) are now commonplace. These factors, operating in an environment of high expectations and ambitious goals, contribute to the potential for depression in physicians over and above the incidence of this illness in the general population.

In 2003 the American Medical Association recognized the importance of depression in physicians with a consensus statement. Although the statement cited a 12.8 percent self-reported prevalence of clinical depression in medical graduates, similar to that in the general population, it would not be surprising if a tendency for doctors to deny the existence of emotional problems led to significant underreporting. More revealing perhaps are reports of a higher rate of suicide in physicians compared to the general population (Center et al. 2003; Schernhammer and Colditz 2004).

Doctors are well acquainted with the cardinal symptoms of depression such as sadness, insomnia and inability to experience pleasure. They may, however, fail to recognize subtle changes in their own behavior that result from depression. A normally unflappable doctor who becomes unduly frustrated by minor inconvenience, intolerant of patient behavior, angered by less than perfect performance of medical staff or continually irritated by the inevitable failings of health care systems may well be experiencing depression.

Few influences affect doctor-patient communication as adversely as depression. It is the natural enemy of empathy and patience. When severe it can impair not only judgment but also intellectual function. As with fatigue, it is key to the recognition of depression in his or her own life for a physician to honestly confront symptoms and behaviors like the ones mentioned above. When you recognize signs of depression in yourself or a colleague, give serious thought to seeking or offering help.

Unfortunately, the cultural factors that attach stigma to mental illness in our society apply with special intensity to physicians. The consensus statement identified a number of other barriers to care for depression among physicians, not the least of which was the fact that 35 percent of doctors do not have a regular source of health care for themselves. Paradoxically, doctors appear to be yet another underserved population when it comes to mental health care. We owe it to ourselves and to our patients to change this situation.

Likes and Dislikes

Have you noticed that you experience positive feelings on seeing some names on your appointment schedule and negative feelings on seeing others? Have you found yourself leaving an encounter with a patient feeling inspired, uplifted or enriched? Conversely, have you ever found yourself looking for an excuse to conclude a visit early or avoid making a follow-up appointment, delaying a hospital call or hoping that a particular patient would decide to see another doctor?

Good doctors tend to resist these behaviors and deny their significance. In our role as physicians we like to think of ourselves as above making subjective judgments about others. A few of us, endowed with an exceptionally generous spirit and a larger than average store of the milk of human kindness, rarely or never experience these feelings and behaviors. Most of us, however, are no different from ordinary mortals and find, sometimes unaccountably, that we like some people and dislike others.

Here I am not referring to overt bias and prejudice. Individuals who are strongly subject to these character flaws seldom choose a career in a patient care specialty. Instead I am referring to a more visceral, instinctive response to another person based largely on unconscious factors. With the crucial exception of romantic feelings, I don't think we have much to worry about in the case of affection for patients we find easy to like. It may be a more painful experience to witness the ravages of illness or death in someone we like very much. It is wise to honestly confront heightened feelings of failure and loss when such a patient dies. But, unless our distress prevents us from advising a difficult but appropriate treatment, we are unlikely to make a bad decision because we like someone. In fact, it is probably true that a bond of genuine affection between doctor and patient exerts a beneficial effect on healing.

On the other hand, some people just rub you the wrong way. If this is a person who has already left the care of a number of other competent doctors before coming to see you, your reaction may not be the result of intolerance on your part. Although your reaction could result from some quirk in your own character it is also possible that you have encountered someone who is genuinely unpleasant, manipulative or even sociopathic. When this happens in ordinary human affairs a sensible strategy is simple avoidance.

For a physician who is responsible for providing medical care for such a person avoidance is not an option and the situation is far from simple. Whatever causes a doctor to respond to a patient with instinctive negative feelings, it has a dreadful influence on the doctor-patient relationship. An unusually strong negative reaction toward a patient will, at a minimum, be a barrier to effective communication. At worst, it will diminish your ability to make sound, objective decisions and set the stage for hard feelings and bad outcomes. So what to do about the patient you just don't like?

First, it is time for another soul-searching personal inventory. Are you, in fact, responding to a previously suppressed bias or prejudice? If so, you are in a position to acknowledge the bias and deal with it rationally. If you succeed in doing this you will have improved your character. Next, is there some specific behavior on your patient's part that is provoking your negative reaction? If you can realistically expect that your patient is capable of altering that behavior -while giving due credit to the fact that this is a sick person who is already dealing with the burden of illness - consider talking about it with the patient.

Explain in a respectful and nonjudgmental way how the patient's actions or inactions are an obstacle to treatment and ask if he or she would make an effort to change in order to help you provide care. For instance, "I have noticed that you [have been missing appointments, are always been late, aren't taking your medicine, haven't bathed in a long time . . .]. If there is some way I can help you [with transportation, reminders, organize your pills, get access to bathing facilities], it would make it easier to manage your care." If you succeed in making the patient your partner in this process you both win.

Or is the actual source of your aversion the fact that you have failed to solve a medical issue? Finding your best efforts unsuccessful in relieving a patient's symptoms, diagnosing the illness or halting the progress of the disease can be enormously frustrating. Unfortunately, these feelings can lead us to blame the patient for our own failure. A large measure of self-awareness is required to recognize that you have fallen victim to this reaction. The healthy response to one's failure to solve a problem is to acknowledge defeat and admit that some problems are insoluble or to ask for help, whether it involves seeking a consultation or referring the patient to another physician.

Finally, are you dealing with an angry or hostile person whose feelings you are reflecting? Or, most difficult of all, have you, in fact, encountered a sociopath? Both situations call for great caution. Such individuals often provoke negative reactions in order to confirm their conviction that others are responsible for their problems. Do not attempt to address anger and hostility directly with a patient in a setting where you are angry, ill at ease or pressed for time. Think carefully about physical security if there is any hint that you might be dealing with a violent or psychotic person.

When it does appear appropriate to talk about anger you may open a discussion by saying, "Something seems to be troubling you. Is it anything I can help to resolve?" Be prepared to sit back and listen. If your patient is willing to describe the source of his or her anger it may be that the grievances are actually reasonable or concern some third party rather than you. In the case of anger directed at a third party it is prudent to give an expression of regret for the patient's experience without being drawn into a discussion of the details.

Your response must be as constructive as possible. If it turns out that the patient's complaint is a reasonable one, you might be able to fix it. In the case of unreasonable grievances or imagined slights, it may be difficult for you to avoid sounding defensive. In all cases, it is vital to do so. It is most important to convey respect for the validity of the patient's feelings, if not the entire substance of the complaint. With luck, listening and giving the patient a chance to ventilate will go a long way toward improving his or her frame of mind.

There remains the rare case where, in spite of efforts to find a source for your feelings or a way to correct the situation, you find it impossible to continue to dealing with a particular individual. You will be faced with two equally undesirable alternatives, either to carry on in spite of your feelings or to announce that you cannot continue to care for the patient. Discharging a patient is an action fraught with potential for adverse consequences, among them liability for a charge of abandonment. This step must always be accompanied by a well-documented referral to a competent and available alternative provider. It is also prudent to seek outside, expert advice before taking action.

All things considered, unless you feel that your safety is at risk, it is wiser to carry on as long as you are able to honestly confront your dislike while taking extra care to behave ethically and professionally to the best of your ability. Facing up to your inability to overcome these feelings is more constructive than looking for ways to excuse or deny them.

This may also help to guard against the impulse to avoid the patient or use other behaviors that might compromise care.

There is a common theme in my approach to dealing with fatigue, depression, likes and dislikes and other human frailties encountered while working in the role of physician. It is embodied in a seldom-used word that is found in the literature of psychiatry from time to time, *autognosis.* Autognosis is defined as self-awareness, the recognition of one's own tendencies, character or peculiarities. It is perilous to consider ourselves immune to human frailty. We are not supermen or superwomen. But if we succeed in exercising the quality of autognosis, we can be cognizant of our limitations and have at least a chance to overcome them. This is a wiser course than maintaining the illusion of perfection.

CHAPTER TEN   CLOSING WORDS

Reflecting on what I have written thus far, I wonder if I have left the impression that there is a single right way to talk to sick people; some formula that, if faithfully followed, guarantees that communication between doctor and patient will answer the needs of the sick - for understanding, lucid information, emotional support and therapeutic advice   This would be a false impression. I have known many fine clinicians who are masters of effective communication and routinely engender feelings of confidence, respect and affection in their patients. No two of these exemplary clinicians use exactly the same style of speaking or choice of words. If they do so well while expressing themselves in different ways, do they share any common characteristics or behaviors that explain their success? They do. These characteristics, displayed by the most effective clinicians, lead the way to effective communication and deserve emphasis.

Understanding

Through experience or instinct, these clinicians have a deep understanding of the ways in which illness changes how people think and act - causing fear in some, determination in others, suspicion and distrust in still others, confidence in others, energizing some, weakening the will of others. In addition understanding these changes, they observe their patients closely enough to recognize which of them are at work in a patient at each visit. The ability to do this allows a clinician to calibrate his or her manner of speaking to adjust to the unique needs of an individual patient. Close observation extends to monitoring a patient's reaction to what you are saying. The use of repetition, further explanation, questions, humor, compassion, silence, praise and a host of other techniques will lead to an optimal response to that reaction.

Honesty

These effective communicators also share the virtue of honesty. A special kind of honesty is required in talking to sick people. It goes beyond simply telling the truth. In fact, in medicine, disclosing all the truth prematurely or gratuitously can be harmful rather than helpful. Instead, these clinicians are able to gauge a patient's capacity to absorb information and carefully meter the release of information to match this capacity.

Honesty in medicine should also encompass sincerity and precision - sincerity in the sense of meaning exactly what you say and precision in the sense of avoiding the platitudes, clichés and casual pleasantries that lubricate conventional small talk. Heightened sensitivity, common among sick people, causes small inconsistencies, careless estimates, euphemisms and evasions in a doctor's speech to have a pernicious influence on patient trust.

Honesty in medicine must also be moderated by compassion. We have to mask our own feelings of sadness and apprehension when talking to patients about painful issues. When unwelcome truth must be faced, many doctors are capable of talking about it without revealing their own emotions. The mark of a truly effective clinician is to be able to do this without sounding cold or uncaring.

Tolerance

Effective communicators are also characterized by an exceptional ability to exercise tolerance. Although most physicians choose their careers, at least in part, because they want to help people, effective communication requires more than a desire to heal. Tolerance is critical, because "sickness very rarely brings to flower the very best of human characteristics" (Tumulty 1973, p. 12). This unfortunate consequence of illness means that, in trying circumstances that demand high-level performance, the clinician is likely to encounter anger, fear, suspicion, dependence or other unhelpful behaviors in a patient.

Such behaviors are understandable and pardonable in the sick. But only a consistently tolerant physician is able to suppress a negative response to difficult behavior and continue to perform at a high level. Doctors who have genuine interest in people, who have also come to know their patients well in less stressful settings, have the best chance of meeting this challenge successfully.

Humility

Humility, in this context, is not to be equated with modesty or a lack of self-confidence. It is not possible to take responsibility for the care of very sick people without a strong belief in your own knowledge and skill. I am referring to a quality that is best understood as the antithesis of arrogance. Arrogance is a trait that will negate the value of even the most brilliant medical advice. Unfortunately, the lengthy and strenuous preparation for a career in clinical medicine sometimes leads to an inflated sense of one's authority and importance. Undo reliance on recent dazzling advances in pharmacology and medical technology also fosters this tendency.

The best defense against arrogance is to remember that there are still many conditions over which we have little control and little to offer except to provide comfort. The most effective clinicians possess the self-confidence required to deal with serious illness without losing the quality of humility.

Communication and the Future of Medicine

I also wonder whether much of what I have written is in danger of becoming obsolete if communication by means of one electronic device or another continues to grow in importance. Many doctors are already adopting email as an efficient way to communicate with their patients. The doctors of future decades are now teenagers. By some accounts they are spending more time exchanging thoughts via email and text messaging than talking to each other. Will they eventually look upon face-to-face verbal communication with the sick as a quaint anachronism - the way we might look at the Norman Rockwell view of a doctor making a home visit, sitting at a sick child's bedside?

The future of medical practice in America has never seemed more fraught with hazard. As I write this, our nation has once again recognized a crisis in health care. Fewer young physicians are planning a career in primary patient care. Fewer still are choosing to practice outside the often rigid structure of large organizations. Our government is engaged in a contentious struggle to find ways to take advantage of the resources science has brought to the practice of medicine without bankrupting the system that delivers these resources to the sick.

Powerful participants in the struggle seem more devoted to protecting their own stake in the ultimate solution than providing a program of just, efficient and uniformly high quality health care. One faction talks about waste, fraud and abuse. Another complains about the effect of malpractice litigation and defensive medicine. Yet another warns about the pernicious influence of profit on hospitals and physicians. Advocates for social justice point to the profit-driven behavior of insurance companies. At the same time worries are growing that advances in our ability to perform interventions in human biology are exceeding our ability to set ethical boundaries.

Relatively little attention has been paid to the role communication could play in resolving some of these problems. Honesty, transparency and better public education relating to health would help to deter waste, fraud and abuse. Effective communication between doctors and patients clearly reduces the probability of malpractice liability. If accurate, evidence-based information about the utility of treatments, diagnostic tests and procedures were properly communicated, profit driven-behavior by providers would become less prevalent. If insurance companies and other healthcare payors could be convinced to recognize the importance of time spent talking to patients and reward it with more appropriate compensation, doctors might devote more time to communication. Public discussion of ethical controversies might lead to a constructive consensus if it were better informed and less passionate.

In addition to caring for the sick, doctors have a potentially powerful role to play in fostering effective communication in each of these areas. I hope my readers will be encouraged to become advocates for this role as well as for better ways of talking to the sick.

Bibliography

Ambady N, Laplante D, Nguyen T, Rosenthal R, Chaumeton N and Levinson W (2002). Surgeon's Tone of Voice: a Clue to Malpractice History. *Surgery*. **132**: 5-9.

Amer A and Fischer H (2009). How Parents Want to Be Greeted By Their Pediatrician. *Clinical Pediatrics*. **48**: 720-22.

Astrow AB, and Sulmasy DP (2004). Spirituality and the Patient-Physician Relationship. *JAMA*. **291**: 2884.

Astrow AB, Wexler A, Texeira K, He MK and Sulmasy DP (2007). Is Failure to Meet Spiritual Needs Associated with Cancer Patients' Perceptions of Quality of Care and Their Satisfaction With Care? *J Clin Oncology*. **25**: 5753-57.

Barrier PA, Li JTC, and Jensen NM (2003). Two Words To Improve Physician-Patient Communication: What Else? *Mayo Clin Proc*. **78**: 211-14.

Beach MC, Roter D, Rubin H, Frankel R, Levinson W and Ford DE (2004). Is Physician Self-Disclosure Related to Patient Evaluation of Office Visits? *J Gen Intern Med*. **19**: 905-10.

Beck RS, Daughtridge R and Sloane PD (2001). Physician-Patient Communication in the Primary Care Office: A Systematic Review. *J Am Board Fam Pract*. **15**: 25-38.

Berlin L (2006). Will Saying "I'm Sorry" Prevent a Malpractice Lawsuit? *Am J Roentgenology*. **187**: 10-15.

Braddock CH and Snyder L (2005). The Doctor Will See You Shortly. The Ethical Significance of Time for the Patient-physician Relationship. *J Gen Intern Med*. **20**: 1057-62.

Brown JB, Boles M, Mullooly JP and Levinson W (1999). Effect Of Clinician Communication Skills Training on Patient Satisfaction. A Randomized, Controlled Trial. *Ann Intern Med.* **131**: 822-29.

Brown RF, Butow PN, Ellis P, Boyle F and Tattersall MH (2004). Seeking Informed Consent to Cancer Clinical trials: Describing Current Practice. *Soc Sci Med.* **58**: 2445-57.

Bruera E, Palmer JL, Pace E, Zhang K, Willey J, Strasser F and Bennett MI (2007). A Randomized, Controlled Trial of Physician Postures When Breaking Bad News. *Palliative Med.* **21**: 501-5.

Buckman R (1992). *How to Break Bad News: A Guide for Health Care Professionals.* Johns Hopkins University Press, Baltimore.

Burton C (2003). Beyond Somatisation: a Review of the Understanding of Medically Unexplained Physical Symptoms (MUPS). *British J Gen Pract.* **53**: 233-41.

Carrol RT (2003). *The Skeptic's Dictionary.* John Wiley & Sons, Hoboken.

Center C et al. (2003). Confronting Depression and Suicide in Physicians. A Consensus Statement. *JAMA.* **289**: 3163-66.

Cherry DK, Woodwell DA and Rechtsteiner EA (2007). National Ambulatory Medical Care Survey: 2005 Summary. *Adv Data.* **387**: 1-39.

Chibnall JT and Brooks CA (2001). Religion in the Clinic: The Role of Physician Beliefs. *South Med J.* **94**: 374-79.

Conill A (2009). Listening Is Powerful Medicine. *National Public Radio Weekend Edition Sunday.* February 1, 2009.

Crowson CS, Therneau TM, Matteson EL and Gabriel SE (2007). Primer: Demystifying Risk: Understanding and Communicating Medical Risks. *Nat Clin Pract Rheumatology.* **3**: 181-87.

Edwards A and Lindsay P (1997). Communication About Risk-Dilemmas for General Practitioners. *Brit J Gen Practice.* **47**: 739-42.

Eggly S, Albrecht TL, Harper FW, Foster T, Franks MM and Ruckdeschel JC (2008). Oncologist' Recommendations of Clinical Trial Participation to Patients. *Patient Educ Counseling.* **70**: 143-48.

Eisenberg DM (1997). Advising Patients Who Seek Alternative Medical Therapies. *Ann Intern Med.* **127**: 61-69.

Ellis MR, Vinson DC, and Ewigman B (1999). Addressing Spiritual Concerns of Patients: Family Physicians' Attitudes and Practices. *J Fam Practice.* **48**: 105-9.

Epstein RM, Alper BS and Quill TE (2004). Communicating Evidence for Participatory Decision Making. *JAMA.* **291**: 2359-66.

Gallagher TH, Garbutt JM, Waterman AD, Flum DR, Larson EB, Waterman BM, Dunagan WC, Fraser VJ and Levinson W (2006). Choosing Your Words Carefully: How Physicians Would Disclose Harmful Medical Errors to Patients. *Arch Intern Med.* **166**: 1585-93.

Gawande A (2007). The Checklist. *The New Yorker.* December 10, 2007.

Gawande A (2009). A Surgical Checklist to Reduce Morbidity and Mortality in a Global Population. *NEJM.* **360**: 491-99.

Gillette RD, Filak A and Thorne C (1992). First Name or Last: Which do Patients Prefer? *J Am Board Fam Pract.* **5**: 517-22

Granek-Catarivas M, Goldstein-Ferber S, Azuri Y, Vinker S and Kahan E (2005). Use of Humor in Primary Care: Different Perceptions Among Patients and Physicians. *Postgrad Med J.* **81**: 126-30.

Halpern J (2007). Empathy and Patient-Physician Conflicts. *J Gen Int Med.* **22**: 696-700.

Haynes AB, Weiser TG, Berry WR, Lipsitz SR, Breizat AH, Dellinger EP, Herbosa T, Joseph S, Kibatala PL, Lapitan MC, Merry AF, Moorthy K, Reznick RK, Taylor G, and Gawande A (2009). A Surgical Safety Checklist to Reduce Morbidity and Mortality in a Global Population. *NEJM.* **360**: 491-99.

Hobma S, Ram P, Muijtjens A, van der Vleuten C and Grol R (2006). Effective Improvement of Doctor-Patient Communication: A Randomized Controlled Trial. *Br J Gen Pract.* **56**: 580-86.

Johnson RL, Sadosty AT, Weaver AL and Goyal DG (2008). To Sit or Not to Sit. *Ann Emerg Med.* **51**: 188-93.

Kachalia A, Shojania KG, Hofer TP, Piotrowski M and Saint S (2003). Does Full Disclosure of Medical Errors Affect Malpractice Liability? The Jury is Still Out. *Jt Comm J Quality and Safety.* **29**: 503-11.

Kaldjian LC, Jones EW, Wu BJ, Forman-Hoffman VL, Levi BH and Rosenthal GE (2007). Disclosing Medical Errors to Patients: Attitudes and Practices of Physicians and Trainees. *J Gen Internal Med.* **22**: 988-96.

King DE and Wells BJ (2003). End-of-Life Issues and Spiritual Histories. *South Med J.* **96**: 391-93.

Knox R, Butow PN, Devine R and Tattersall MH (2002). Audiotapes of Oncology Consultations: Only for the First Consultation? *Ann Oncol.* **13**: 622-27.

Koenig HG (2004). Taking a Spiritual History. *JAMA.* **291**: 2881.

Kraman SS and Hamm G (1999). Risk Management: Extreme Honesty May be the Best Policy. *Ann Intern Med.* **131**: 963-67.

Levinson W, Roter DL, Mullooly JP, Dull VT and Frankel RM (1997) Physician-patient Communication. The Relationship with Malpractice Claims Among Primary Care Physicians and Surgeons. *JAMA.* **277**: 553-59.

Lin CT, Albertson GA, Schilling LM, Cyran EM, Anderson SN, Ware L, Anderson RJ (2001). Is Patients' Perception of Time Spent With the Physician a Determinant of Ambulatory Patient Satisfaction? *Arch Intern Med.* **161**: 1437-42.

Liddell C, Rae G, Brown TR, Johnston D, Coates V and Mallett J (2004). Giving Patients an Audiotape of Their GP Consultation: A randomized Controlled Trial. *Br J Gen Pract.* **54**: 667-72.

Maes M (2009). Inflammatory and Oxidative and Nitrosative Stress Pathways Underpinning Chronic Fatigue, Somatization and Psychosomatic Symptoms. *Curr Opin Psychiatry.* **22**: 75-83.

Makoul G, Zick MA and Green M (2007). An Evidence-Based Perspective on Greetings in Medical Encounters. *Arch Intern Med.* **167**: 1172-76.

Mathews SC, Camacho A, Mills PJ and Dimsdale JE (2003). The Internet for Medical Information About Cancer: Help or hindrance? *Psychosomatics.* **44**: 100-3.

McDaniel S, Beckman HB, Morse DS, Silberman J, Seaburn DB and Eps (2007). Physician Self-Disclosure in Primary Care Visits. *Arch Intern Med.* **167**: 1321-26.

McDonald IG, Daly J, Jelinek VM, Panetta F, and Guttman JM (1996). Opening Pandora's Box: The Unpredictability of Reassurance by a Normal Test Result. *BMJ.* **313**: 329-32.

McLafferty RB, Williams RG, Lambert AD and Dunnington GL (2006). Surgeon Communication Behaviors That Lead Patients to Not Recommend the Surgeon to Family Members or Friends: Analysis and Impact. *Surgery.* **140**: 622-24.

Mazor KM, Simon SR and Gurwitz JH (2004). Communicating With Patients About Medical Errors: A Review of the Literature. *Arch Intern Med.* **164**: 1690-97.

Mechanic D, McAlpine DD and Rosenthal M (2001). Are Patients' Office Visits With Physicians Getting Shorter? *NEJM.* **344**: 198-204.

Merckaert I, Libert Y, Delvaux N, Marchal S, Boniver J, Etienne AM, Klastersky J, Reynaert C, Scalliet P, Slachmuylder JL and Razavi D (2005). Factors That Influence Physicians' Detection of Distress in Patients With Cancer: Can a Communication Skills Training Program Improve Physicians' Detection? *Cancer.* **104**: 411-21.

Meit SS, Williams D, Mencken FC and Yasek V (1997). Gowning: Effects on Patient Satisfaction. *J Fam Pract.* **45**: 397-401.

Monroe MH, Bynum D, Susi B, Schultz L, Franco M, MacLean CD, Cykert S and Garrett J (2003). Primary Care Physician Preferences Regarding Spiritual Behavior in Medical Practice. *Arch Intern Med.* **163** 2751-56.

Morgan W and Engel G (1969). *The Clinical Approach to the Patient.* W. B. Saunders, Philadelphia.

Meyers JL, Moore C, McGrory A, Sparr J and Ahern M (2004). Use of the Physician Orders for Life-Sustaining Treatment (POLST). *J Gerontological Nursing.* **30**: 37-46.

Noyes R Jr., Stuart SP and Watson DB (2008). A Reconceptualization of the Somatoform Disorders. *Pyschosomatics* **49**: 14-22.

Page LA and Wessely S (2003). Medically Unexplained Symptoms: Exacerbating factors in the Doctor-patient Encounter. *J R Soc Med.* **96**: 223-27.

Pelt JL and Faldmo LP (2008). Physician Error and Disclosure. *Clin Obstet Gyncol.* **51**: 700-8.

Petrie KL, Müller JT, Schirmbeck F, Donkin L, Broadbent E, Ellis CJ, Gamble G and Rief W (2007). Effect of Providing Information About Normal Test Results on Patients' Reassurance: Randomized Controlled Trial. *BMJ.* **334** :352-55.

Rhoades DR, McFarland KF, Finch WH and Johnson AO (2001). Speaking and Interruptions During Primary Care Office Visits. *Fam Med.* **33**: 528-32.

Schernhammer ES and Colditz GA (2004). Suicide Rates Among Physicians: A Quantitative and Gender Assessment (meta-analysis). *Am J Psychiatry.* **162**: 2295-302.

Schneider AP, Nelson DJ and Brown DD (1993). In-Hospital Cardiopulmonary Resuscitation: A 30-year Review. *J Am Board Fam Pract.* **6**: 91-101.

Silberman J, Tentler A, Ramgopal R and Epstein RM (2008). Recall-Promoting Physician Behaviors in Primary Care. *J Gen Intern Med.* **23**: 1487-90.

Sontag S (1978). *Illness as Metaphor* Farrar, Straus & Giroux, New York.

Stone J, Smyth R, Carson A, Lewis S, Prescott R, Warlow C and Sharpe M (2005). Systematic Review of Misdiagnosis of Conversion Symptoms and "Hysteria." *BMJ*. **331**: 989-94.

Tervalon M and Murray-Garcia J (1998). Cultural Humility Versus Cultural Competence: A Critical Distinction in Defining physician Training Outcomes in Multicultural Education. *J Health Care for the Poor and Underserved*. **9**: 117-25.

Trillin AS (2001) Betting Your Life. *The New Yorker*. January 29, 2001.

Tumulty P (1973). *The Effective Clinician*. W. B. Saunders, Philadelphia.

Ulene V (2009). Bad News, Bad Delivery. *Los Angeles Times* March 9, 2009.

Yedidia MJ, Gillespie DD, Kachur E, Schwartz MD, Ockene J, Chepaitis AE, Snyder CW, Lazare A and Lipkin M Jr (2003). Effect of Communications Training on Medical Student Performance. *JAMA*. **290**: 1210-12.

Index